The Physics and Mathematics of Adiabatic Shear Bands

This book is a research monograph on the material instability known as adiabatic shear banding, which often occurs in a plastically deforming material as it undergoes rapid shearing. Plastic deformation generates heat, which eventually softens most materials with continued straining, a process which is usually unstable. In this case the instability results in thin regions of highly deformed material, which are often the sites of further damage and complete failure. Divided into three parts, the book first reviews the physical phenomenon and the standard methods of testing and characterization. It then establishes a general theory of isotropic plasticity with finite deformations as a setting for the simpler, but still nonlinear and highly coupled, equations of adiabatic shearing and the idealizations that are necessary to establish them. The main body of the book examines a series of one-dimensional problems of increasing complexity. In this way a comprehensive and quantitative picture of the complete phenomena is built up. Particular care is taken to use well-established asymptotic techniques to find simple, but universal, analytic expressions or scaling laws that encapsulate various aspects of the dynamic formation and the final morphology of shear bands. The last two chapters review recent two-dimensional experiments and analyses.

A fully developed mechanics of shear is just beginning to emerge as a major companion to fracture mechanics; this book may speed the process along.

Dr. T.W. Wright is a Senior Research Scientist at the U.S. Army Research Laboratory at Aberdeen Proving Ground, MD, where he has worked since 1967. He has published more than 60 papers, most recently in the *Journal of Mechanics and Physics of Solids*, *International Journal of Plasticity*, *Mechanics of Materials*, *International Journal of Solids and Structures*, and *Acta Materialia*. Dr. Wright is a Fellow of the American Society of Mechanical Engineers and of the American Academy of Mechanics. He currently serves on the Editorial Board of the *International Journal of Plasticity*.

OTHER TITLES IN THIS SERIES

THE PHYSICS AND MATHEMATICS OF ADIABATIC SHEAR BANDS

T.W. WRIGHT

Army Research Laboratory

PUBLISHED BY THE PRESS SYNDICATE OF THE UNIVERSITY OF CAMBRIDGE
The Pitt Building, Trumpington Street, Cambridge, United Kingdom

CAMBRIDGE UNIVERSITY PRESS
The Edinburgh Building, Cambridge CB2 2RU, UK
40 West 20th Street, New York, NY 10011-4211, USA
477 Williamstown Road, Port Melbourne, VIC 3207, Australia
Ruiz de Alarcón 13, 28014 Madrid, Spain
Dock House, The Waterfront, Cape Town 8001, South Africa

http:// www.cambridge.org

First published 2002

Printed in the United Kingdom at the University Press, Cambridge

Typeface Times Roman 10/13 pt.   *System* LATEX 2$_\varepsilon$ [TB]

A catalog record for this book is available from the British Library.

Library of Congress Cataloging in Publication Data
Wright, T.W.
The physics and mathematics of adiabatic shear bands / by T.W. Wright.
p.   cm. – (Cambridge monographs on mechanics)
Includes bibliographical references and index.
ISBN 0-521-63195-5
1. Shear (Mechanics)   2. Flexure.   3. Materials – Thermal properties.
I. Title.   II. Series.
TA418.7.S5 W75   2002
620.1′1245 – dc21   2001043623

ISBN 0 521 63195 5 hardback

Dedicated with much love to Margaret, Andy, Charlie, and Kate

Contents

Preface

In the past decade or so there have been several reviews on the topic of adiabatic shear bands in the engineering literature; there has also been a full-length book by Bai and Dodd (1992). The present book is not intended to be a review, but rather it attempts a coherent exposition of the mechanics of band formation, presented for the most part from a single point of view. It differs from the book by Bai and Dodd by adopting a more mathematical approach and by concentrating on special solutions as a way both of elucidating dynamic behavior and of developing scaling laws that summarize that behavior. In fact, development of scaling laws (simple formulas that connect the dynamical environment, the morphological structure, and the physical and mechanical properties) is a central theme of the book. Other themes that appear throughout the book are the interplay between numerical and analytical approaches and the more or less systematic use of asymptotic techniques to extract simple descriptions of complex, nonlinear processes. Much of the material is based on lectures given originally when the author was a short-term visitor at the University of California at San Diego in January 1990, later expanded, and given again at the university in February 1995.

The introductory chapter has two parts. The first makes generous use of pictures and photomicrographs to give the reader a general sense of the scale and morphology of shear bands. At the same time, the reader should gain an impression of which metals form bands and under what dynamic circumstances the bands form. There is also a discussion of the implications that the small, localized scale of shear bands has for numerical simulations. The second half of the chapter reviews selected parts of the experimental literature, first covering general dynamic testing to obtain material properties, and then specialized testing to characterize adiabatic shear bands. The latter experiments show the typical sequence of band formation; that is, uniform work hardening and heating from plastic work, initial slow growth of perturbations, attainment of maximum stress followed by slow decay of stress for an indeterminate length of time, sudden acceleration of local heating and strain rate accompanied by a loss of stability and the capacity to transfer shear stresses, and culmination in a fully formed band that is nearly a line source (or more accurately, a surface source) of heating. The experiment that illustrates this behavior most clearly is the torsional Hopkinson bar test, and some of the chapter is devoted to a brief description of that experiment and of some typical data.

Chapters 2 and 3 review the governing equations – continuum balance laws, nonlinear thermoelasticity and plasticity, and the structure of isotropic flow laws – with careful attention paid to simplifying assumptions and specializations, especially as applied to the usual one-dimensional equations for adiabatic shear bands. In the course of that development it becomes apparent that there is still room for improvement in certain aspects of the constitutive laws for plastic work and heating from plastic work. Chapter 4 gives a phenomenological treatment of work hardening, followed by brief descriptions of seven typical flow models for thermoviscoplasticity. These first four chapters as a group set the physical and mathematical stage for detailed study of adiabatic shearing.

Chapters 5, 6, and 7 explore the dynamics of band formation in a one-dimensional setting. Examination of a sequence of increasingly complex cases reveals most of the major features of band formation. These features include the timing of extreme localization, the morphology of fully developed bands, and the quantitative role of various physical properties such as thermal conductivity, heat capacity, work hardening, thermal softening, and strain rate sensitivity. Wherever possible the results are summarized in the form of simple scaling laws so as to be as useful and accessible as possible. Chapter 5 deals with general considerations concerning the development of homogeneously softening regimes and their potential instability. Without heat conduction and strain rate effects, the dynamic governing equations in one dimension may show a change of type from hyperbolic in space and time to elliptic, which for initial value problems indicates a change from wave propagation to localization and ultimately singular behavior. Strain rate sensitivity has the effect of delaying, but not eliminating, the singularity, and thermal conductivity has the effect of regularizing the equations so that although the solution may still localize strongly, the singularity itself is avoided. Other theoretical approaches postulate an intrinsic, material length scale that also has the effect of regularizing the equations and thus limiting the extent of localization, but in the present treatment the limiting length scale arises naturally from the competition between heat conduction and heat generation at a strong, self-powered source. Chapter 6 explores the initial stages of instability and the early growth of perturbations, but even though the equations used here are linear, it is possible to extract important scaling parameters and to estimate the timing of full localization in the form of scaling laws. Chapter 7 explores the fully nonlinear equations directly. Even with work hardening included, exact solutions (albeit in intrinsic and parametric form) may be found for a few special cases if heat conduction is eliminated. If heat conduction is included, there appears to be only one exact solution known, and that for a rather special case. However, approximate solutions are available, and from these, better estimates of the timing of full localization can be made. Indeed, in the absence of work hardening, a universal approximate solution is available for arbitrary thermal softening and arbitrary, but weak, strain rate sensitivity. The asymptotic structure of a fully formed band is also examined in Chapter 7.

The last two chapters extend the discussion to two dimensions. Chapter 8 summarizes some of the principal results from three kinds of two-dimensional experiments. The first, called the Kalthoff experiment, produces a shear band that extends from the tip of a preexisting edge crack in a plate by forcing mode II motion. By applying dynamic torsion, the second type of experiment forces mode III motion in a thick-walled cylinder and drives a shear band radially inward. The third type, called the thick-walled cylinder experiment, produces a spiral array of shear bands in a hollow cylinder that is imploded and crushed by a low-velocity explosive. The first experiment permits direct examination of the region near the tip of a propagating shear band. The second permits only indirect, postmortem examination, but the technique permits the loading pulse to be carefully measured and controlled. The third experiment permits the study of the collective behavior of shear bands and their interaction as they progress toward total failure of the material, but it too permits only postmortem examination.

Propagation is a fundamental dynamical characteristic, which for a shear band must be related to energy dissipation and shear toughness in the material, and the collective behavior must have a strong bearing on the scale of patterns of failure by massive fragmentation. Unfortunately these aspects remain little explored or understood. The final chapter discusses the principal known solutions in two dimensions for propagating shear bands, but the results are still fragmentary, at best, because they are confined to local analyses near the tip of a propagating shear band or to boundary layer and similarity solutions. In these cases, analysis can be guided to some extent by known results in one dimension or by known successful techniques from fluid mechanics.

Although much progress has been made in both theoretical and experimental aspects concerning the formation, characterization, and properties of adiabatic shear bands, a fully developed mechanics of shear has yet to appear. However, with continued research it should develop sufficiently so as to take its place as a major companion to fracture mechanics.

T.W. Wright
Baltimore, MD
March 2001

Acknowledgments

It is my great pleasure to acknowledge the encouragement received from a number of colleagues and friends. Sia Nemat-Nasser, besides acting as host on the occasion of my many short-term visits to the University of California at San Diego (UCSD), suggested that I should write this book at just about the time that I had seriously begun to consider it myself.

Many people influenced the formation of my views and approach to the subject of adiabatic shear bands. The early work of Don Curran, Don Shockey, and Lynn Seaman at SRI International first introduced me to the topic and its prevalence in impact mechanics. Romesh Batra, now at Virginia Polytechnic Institute & State University, spent a year at the Ballistic Research Laboratory, and together we began our study of adiabatic shear. The numerical work of my former colleague, John Walter, now at Los Alamos National Laboratory, was invaluable in illuminating the nature and dynamics of band formation. Eric Varley at Lehigh University introduced me to implicit and parametric ways to attack nonlinear problems. During my fellowship year at Oxford University, Hillary Ockendon demonstrated to me the power of linear techniques for identifying fundamental patterns in nonlinear problems. On several occasions G. Ravichandran made arrangements for me to make short-term visits to Cal Tech. Ravi's generous sharing of his experimental insights was invaluable. Vitali Nesterenko and Marc Meyers at UCSD introduced me to the notion of a self-organized array of shear bands. Others whose work has contributed substantially to my understanding of adiabatic shear bands are Michael Ortiz and Ares Rosakis at Cal Tech, Jim Glimm and Brad Plohr at State University of New York at Stony Brook, and Dave Benson and Ken Vecchio at UCSD.

Several friends and colleagues have been invaluable in reading and critiquing various chapters. K.T. Ramesh at Johns Hopkins and Datta Dandekar at the U.S. Army Research Laboratory (ARL) reviewed Chapter 1 and were of great assistance in straightening out my views on experimental work. Mike Scheidler at ARL read Chapters 2 and 3 with his characteristic thoroughness and helped me to tighten my arguments and presentation in many places. Tracy Vogler at ARL reviewed Chapter 4, Scott Schoenfeld at ARL critiqued Chapters 5, 8, and 9, and Jim Cox at ARL made many useful comments on Chapters 6 and 7. I thank them all for their assistance. Of course, any inaccuracies that remain in the presentation are my responsibility alone.

I am grateful to Addison Wesley Longman, the U.S. Army Research Laboratory, American Society of Metals International, American Society of Mechanical Engineers International, Elsevier Science, the Institute of Physics, Kluwer/Plenum Publishers, Metal Powder Industries Federation/APMI International, Sage Publications, Inc., The Minerals, Metals, and Materials Society (TMS), and Marc A. Meyers for permission to reproduce figures for which they hold the copyrights. Special thanks must be given to Brenda Conjour and Shirleen Ellis of ARL for their always cheerful and careful preparation of the graphics in this book. Again, any difficulties that may remain are due only to me.

The U.S. Army Research Laboratory and the U.S. Army Ballistic Research Laboratory before it have been fully supportive of my work over my many years here as an employee, and I gratefully acknowledge the congenial working environment, knowledgeable and helpful colleagues, and the opportunities for scientific travel that have been provided for me. Diane Parks, Pam Brown, and Kim Aguirre all provided expert secretarial support during the preparation of this manuscript.

Last of all I must express my profound gratitude to Professors Clifford A. Truesdell, deceased, and Jerald L. Ericksen, retired, both formerly at The Johns Hopkins University, with whom I was privileged to spend a few years in the same academic department more than thirty years ago. Although I was never the student of either, it is from their teaching that I learned most of what I know about the foundations of modern continuum mechanics and gained some sense of its reach and power.

1

Introduction: Qualitative description and one-dimensional experiments

Through words and pictures the first half of this chapter defines what the phrase "adiabatic shear band" means. It gives a qualitative description of the appearance and occurrence of adiabatic shear bands, and by simple inference it also gives some rudimentary idea of the typical mechanical and thermal history that accompanies their formation. A discussion of their importance as damage sites and a qualitative description of the basic mechanics that drives them follows. Then, after a brief review of earlier work including some recent surveys and symposiums, Section 1.1 closes with a short discussion of the implications that adiabatic shear bands have for the predictive accuracy of numerical analysis and computational design.

The second half of the chapter gives an overview of the techniques and typical results that experimentalists have developed in trying to gain a more quantitative picture of the phenomenon. For the most part, these techniques have involved ingenious use of the well-known split Hopkinson pressure bar (or Kolsky bar), but a limited amount of work has also been done by other means. There are really two parts to this work. First it is necessary to explore the fundamental constitutive behavior of a given material at high strain rates and various temperatures. Only then does it make sense to measure the forces, temperatures, strains, and so on that develop during the formation of an adiabatic shear band. This is true because the formation of a shear band is merely the consequence of locally evolving material response to a particular, externally imposed environment after homogeneous deformation has become unstable.

There is a theoretical counterpart to this experimental approach. The fundamentals of large plastic deformation, including a reasonably coherent constitutive theory, ought to be understood before the concrete initial-boundary value problems of shear localization can be most fruitfully explored. The remainder of the book is organized with this last thought in mind.

1.1 Qualitative Features of Adiabatic Shear Bands

1.1.1 Basic Morphology

An adiabatic shear band is a narrow, nearly planar or two-dimensional region of very large shearing that sometimes occurs in metals and alloys as they experience intense dynamic loading. Once the band is fully formed, the two sides of the region are displaced relative to each other, much like a mode II or mode III crack, but the material still retains full physical continuity from one side to the other. The thickness of the most heavily sheared region may be of the order of only a few tens of micrometers or even less, whereas the lateral extent may be many millimeters or even centimeters in length. Thus the aspect ratio of the sheared region may be in the hundreds or even thousands. Figure 1.1 shows a typical example where a band has been driven into a 1018 cold-rolled steel specimen by a flat-ended projectile. The picture shows a cross section taken perpendicular to both the bottom and vertical side of the cylindrical crater. Note the long, narrow appearance of the band, beginning at a corner of the crater where a stress concentration was introduced by the impact, and extending deep into the specimen in the direction of impact. Figure 1.2 shows a band, also introduced by impact of a projectile, in a 2014-T6 Al alloy. At the scales shown the internal structure of the fully formed bands themselves cannot be seen in either micrograph. In both cases, however, the intense shear in the surrounding material can be seen, achieving shear strains much greater than one in both

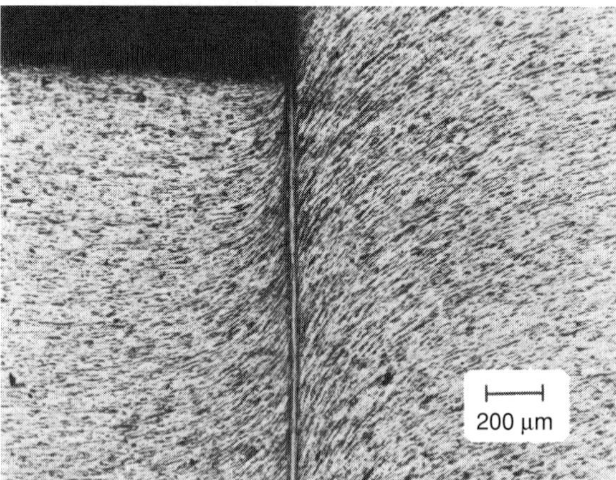

Figure 1.1. Adiabatic shear band in 1018 cold-rolled steel. Band formed at the corner of an impact crater. Impact speed, 100 m/s. (Reproduced from Rogers 1983, with permission from Kluwer Academic/Plenum Publishers.)

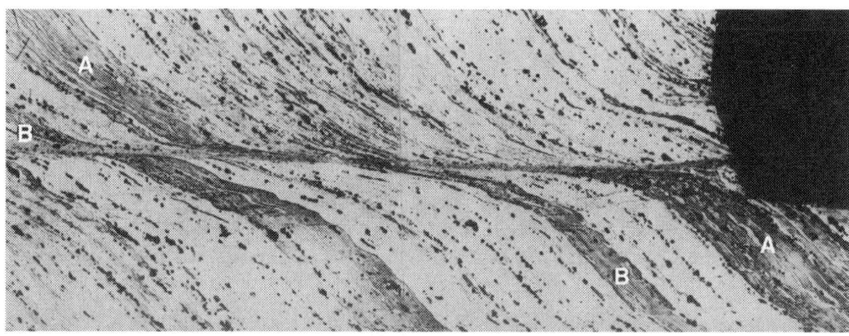

Figure 1.2. Adiabatic shear band in 2014-T6 Al alloy. Band formed at the corner of an impact crater. (Reproduced from Wingrove 1973. Copyright 1973 by the Minerals, Metals and Materials Society and ASM International.)

cases (as estimated by the tangent of the flow lines). In addition, Figure 1.2 gives the impression of very large deformation within the band. A comparison of the relative positions of the dark zones on either side of the band, as indicated by the letters A and B, suggests a relative displacement of perhaps twenty to thirty times the apparent width of the band itself. Accordingly, the maximum shear strain within the band must be substantially greater than ten by even the crudest estimate. Figure 1.3 shows a closer view of an adiabatic shear band that was produced by a punch driven explosively through a rolled, high strength, Ni-Cr steel alloy. The lines that are visible in the photomicrograph are chemical inhomogeneities that lay approximately in the rolling planes, which originally were horizontal in the picture. In this example, the shear strain exterior to the band proper is approximately one, and again, by direct examination of the slope of the marker lines, the strain within the band is greater than ten.

These three examples have introduced several of the principal characteristics of adiabatic shear bands: thinness perpendicular to the direction of maximum shearing, high aspect ratio, high external strain, extreme internal strain, and dynamic generation, the latter inferred from projectile impact and punching, and in turn implying generation under high strain rates. In each of the three cases, the band cuts straight through the material, apparently without regard to the microstructure. This observation seems to hold true in the vast majority of cases, but branching is sometimes observed also. In addition, we observe that bands can form in different materials and that their appearance may be different in different materials.

This last point has been noted in the older literature by a referral to deformed and transformed bands. Transformed bands, like the two steels, show white after etching in preparation for photomicroscopy, whereas deformed bands, like the aluminum alloy, do not. It has often been speculated that a phase transformation

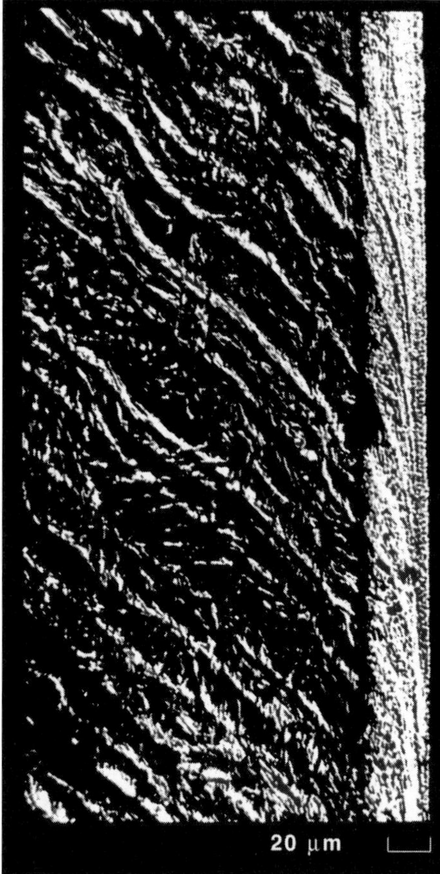

Figure 1.3. Adiabatic shear band in a Ni-Cr steel alloy. Band formed from a punch driven explosively through the specimen. White lines are bands of chemical inhomogeneity and were horizontal before punching. (Reproduced from Moss 1981, with permission from Kluwer Academic/Plenum Publishers.)

has occurred in a transformed band, but at least in some cases this does not appear to be so. Beatty, Meyer, Meyers, and Nemat-Nasser (1992) showed that the center of a shear band in a 4340 steel (RC52 hardness before shearing) contained very fine grains (8–20 nm), probably as a result of grain refinement caused by the severe deformation, but no indication of a transformation from martensite to austenite was apparent. Meunier, Roux, and Moureaud (1992) obtained similar results for an armor steel, as did Chichili (1997) for alpha Ti. In any event, when the internal structure of the band is visible, as in Figure 1.3, the lines of deformation appear to extend more or less smoothly and continuously

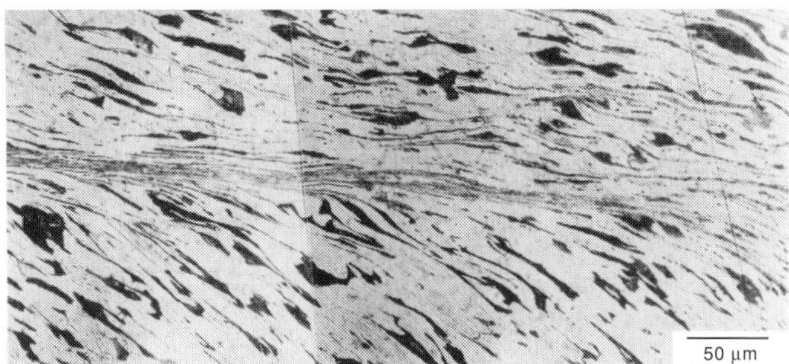

Figure 1.4. Tip of an adiabatic shear band showing the transition from intense localization to a more diffuse deformation. (Reproduced from Meyers and Wittman 1990.
Copyright 1990 by the Minerals, Metals and Materials Society and ASM International.)

through the white etching region, suggesting that even if a phase transformation
has occurred, the kinetics are such that it has not affected the basic processes
of deformation.

After deformation has stopped, however, Figure 1.4 indicates that the end or
tip of a shear band simply becomes diffuse as the intense deformation within the
band transitions into the general deformation of surrounding material. In other
cases, as in Figure 1.5, there is no obvious deformation pattern through the band.
Here there is a rough appearance, as if many smaller grains had been produced
by the thermomechanical deformation or perhaps by recovery processes. Again
the kinetics and timing of the transformation are uncertain.

Figure 1.6 shows a typical microhardness profile across a shear band. Usually the center of the band shows greatly increased hardness, as in this case,

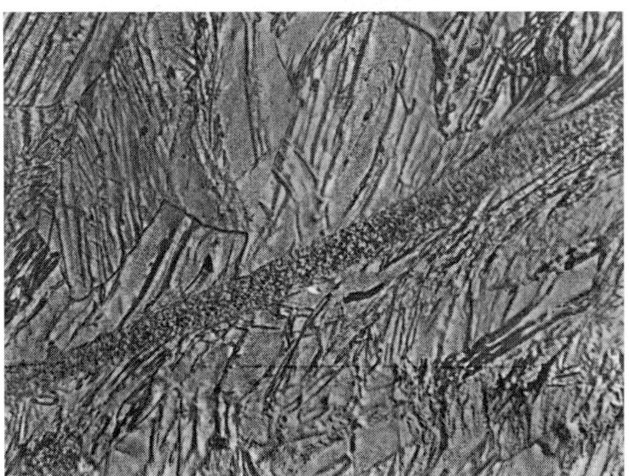

Figure 1.5. Adiabatic shear band in Ti. Material appears to have recrystallized after
band formation. (Reproduced from Grebe, Pak, and Meyers 1985. Copyright by the
Minerals, Metals and Materials Society and ASM International.)

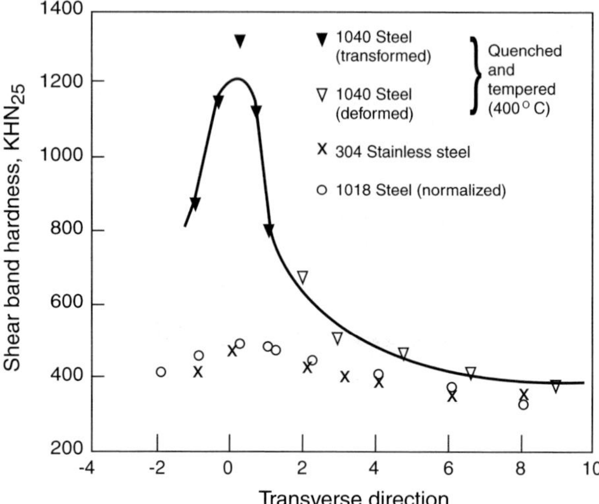

Figure 1.6. Microhardness profiles across shear bands in several steels. (Reproduced from Rogers and Shastry 1981, with permission from Kluwer Academic/Plenum Publishers.)

which might occur from a cycle of work hardening and heating, followed by rapid quenching. Occasionally the profile shows several peaks in hardness, as in Figure 1.7. Besides the morphological characteristics listed in the last paragraph, then there also are indications that somewhat elevated temperatures have occurred within the fully formed band.

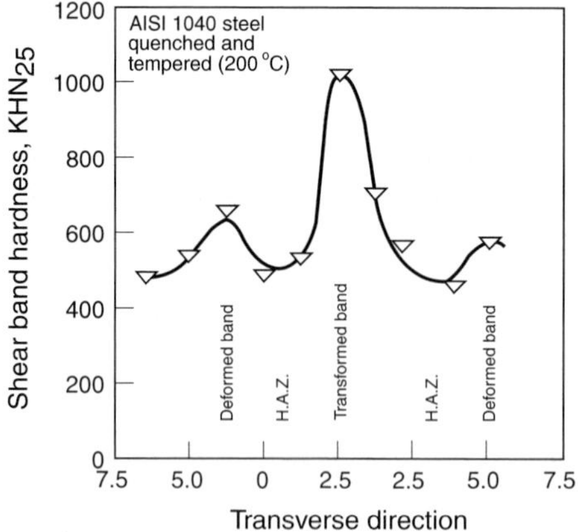

Figure 1.7. Microhardness profile across a shear band in AISI 1040 steel showing a heat-affected zone. (Reproduced from Rogers and Shastry 1981, with permission from Kluwer Academic/Plenum Publishers.)

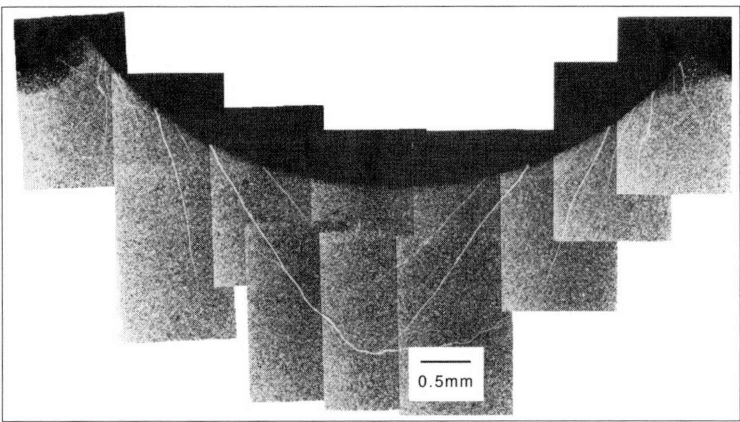

Figure 1.8. Shear bands in Ti-6Al-4V formed below a crater, which was formed by the impact of a steel ball (6.35 mm in diameter) striking at 318 m/s. (Reproduced from Timothy and Hutchings 1985, with permission from Elsevier Science.)

1.1.2 Occurrence

Adiabatic shear bands have been observed to form in many circumstances. Perhaps the most well-known example is ballistic impact, in which adiabatic shear has long been observed in targets and more recently in penetrators as well. Timothy and Hutchings (1985) reported shear bands that formed in the bottom of a crater formed by the impact of a hard steel ball on a target material of Ti-6Aℓ-4V (see Figure 1.8). This study shows that shear bands can form in regions that are free from obvious stress concentrations, such as those examined by Rogers (1979) in his studies. Traditionally shear bands have been regarded as deleterious to the performance of either penetrators or armors, as in projectile shatter or target plugging. More recently, however, Magness (1994) proposed that adiabatic shear can be a beneficial mechanism that enhances the superior performance of U3/4Ti as a penetrator material, even beyond that expected from its high density. The idea is that adiabatic shear failure, occurring at an optimum strain, allows the penetrator material to cease plastic flow and, by failing, to be able to exit the zone of interaction with the target at a time earlier than would have been possible by plastic flow alone. The net result is that the cavity forced into the target is narrower and deeper than would otherwise have been the case. Figure 1.9 is a sketch of the proposed mechanism, which is now widely accepted as the correct explanation. Figure 1.10 shows the characteristic shape of a penetrator in free flight after perforating a target in which adiabatic shear has allowed the penetrator to be "self-sharpening."

Targets sometimes show unusual shear failures. Figure 1.11 shows an example in which the target material tends to fail on surfaces more or less parallel

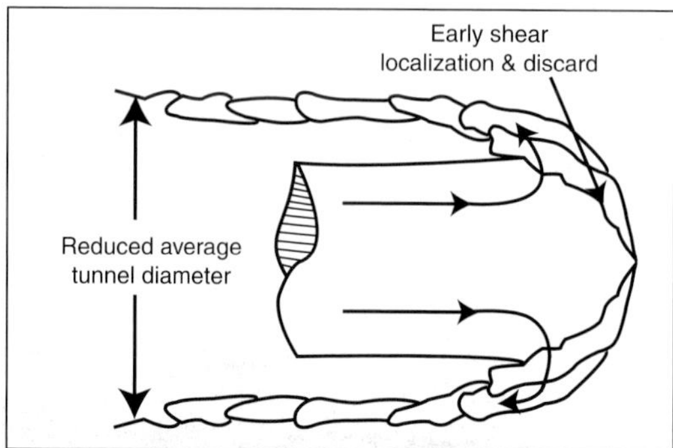

Figure 1.9. Sketch of flow and failure by adiabatic shear in the tip of a penetrator as it penetrates a target. (Mechanism proposed by Magness 1993; reproduced with permission from Metal Powder Industries Federation/APMI International.)

to the path of penetration. Notice the roughness of the failure zone where adiabatic shear has apparently allowed target material to break away intermittently as the penetrator advanced. In this case the total cavity is much wider than would be expected from gross plastic flow alone. In fact, shear failure seems to occur near surfaces of maximum strain rate, as calculated by Batra and Wright (1986). Figure 1.12 shows an example in which shear zones have formed ahead of the advancing penetrator, much like bow waves, but because the shearing

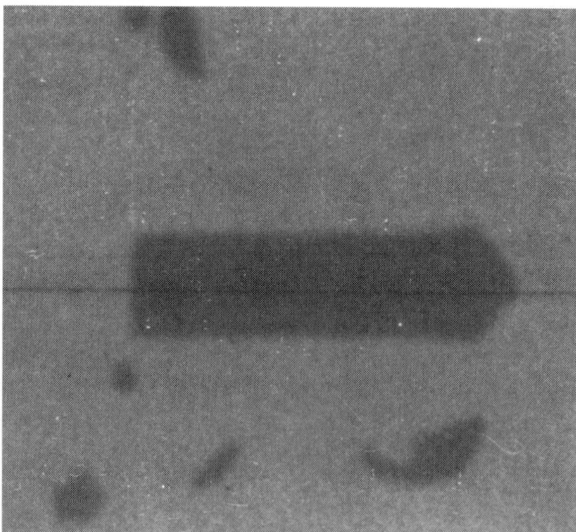

Figure 1.10. Flash x-ray of a U-8Mo penetrator after perforation of a steel target. The chiseled nose is evidence that adiabatic shear has played a strong role in eroding the penetrator. (Reproduced from Magness 1994, with permission from Elsevier Science.)

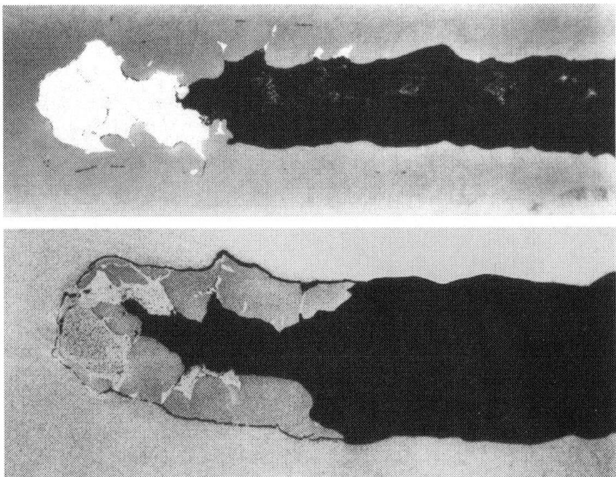

Figure 1.11. Tunnels formed in targets made of Ti-6Al-4V after being struck by a small rod made of 90W-7Ni-3Fe at 1450 m/s. Top figure shows incipient shear parallel to the tunnel at a distance that corresponds to the maximum gradient in the velocity field. Bottom figure shows fully developed shear. (G. E. Hauver, unpublished photographs, with permission from the Terminal Effects Division of the Army Research Laboratory.)

takes place on material surfaces, they cannot advance with the penetrator and so are perforated as penetration continues. New surfaces form and are perforated in sequence until finally penetration stops.

There are also many industrial processes in which adiabatic shear bands may form (Semiatin, Lahoti, and Oh 1983). Figure 1.13 shows a segmented chip formed during machining of Ti-6Aℓ-4V, where the segments are separated by adiabatic shear bands. The formation of such chips places a limit on the quality of surface finish that can be achieved. Additionally, the figure shows

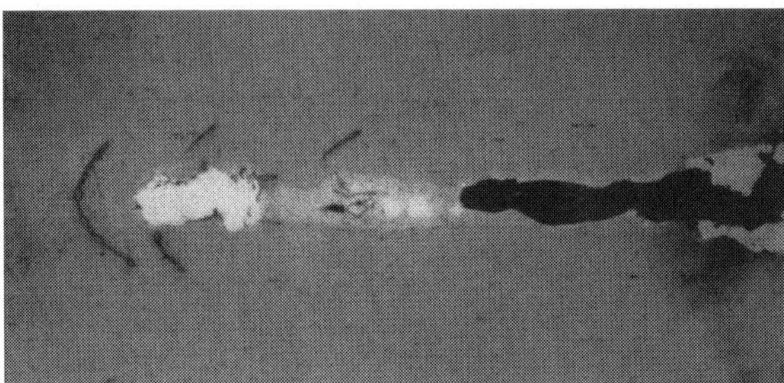

Figure 1.12. Tunnel formed in a target made of S7 tool steel at RC30 after being struck by a small rod made of 90W-7Ni-3Fe at 1450 m/s. Figure shows a sequence of sheared regions that form intermittently ahead of the penetrator, like bow waves. (G. E. Hauver, unpublished photographs, with permission from the Terminal Effects Division of the Army Research Laboratory.)

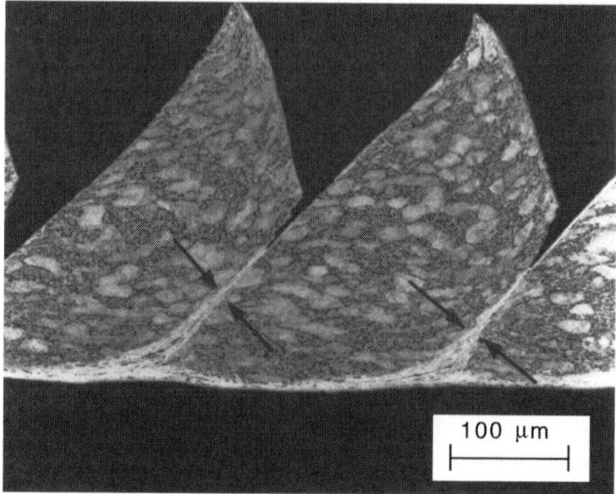

Figure 1.13. Segmented maching chip, formed in Ti-6Al-4V. Segments are separated by adiabatic shear bands. (Reproduced from Semiatin et al. 1983, with permission from Kluwer Academic/Plenum Publishers.)

that shear bands can form and reform intermittently if the loading conditions are maintained, as also seems to be the case in Figures 1.9, 1.11, and 1.12. Forging or upsetting can also produce unwanted adiabatic shear bands. Figure 1.14 shows a cross-shaped region in the middle of the specimen caused by side pressing (Semiatin et al. 1983). A high rate compression has been shown to produce adiabatic shear bands in some materials. Wulf (1978) found shear bands in an Al alloy after compression testing in a split Hopkinson pressure bar; see Figure 1.15. Chen and Meyers (personal communication, but see also Chen, Meyers, and Nesterenko 1999 for further details) found that they could produce shear bands in Ta with high rate compression at 77 K, Figure 1.16, but not at room temperature. Grady, Asay, Rohde, and Wise (1983) observed a cross-hatched deformation pattern in Al after shock impact (Figure 1.17), which they interpreted as possibly being adiabatic shear bands. These examples

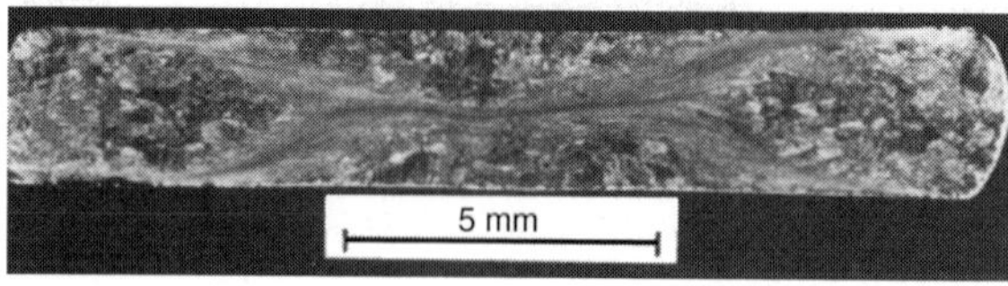

Figure 1.14. Transverse section of a Ti-6242Si cylinder sidepressed at 843 °C. (Reproduced from Semiatin et al. 1983, with permission from Kluwer Academic/Plenum Publishers.)

Figure 1.15. Adiabatic shear band formed in a compression specimen of 7039 Al at a strain rate of $8 \times 10^3 \, \mathrm{s}^{-1}$. Total strain is 0.26. (Reproduced from Wulf 1978, with permission from Elsevier Science.)

demonstrate that adiabatic shear bands may form in many materials during many kinds of deformation.

1.1.3 Importance

Their mere existence would not make adiabatic shear bands worthy candidates for extensive study if they did not significantly alter the subsequent behavior and performance of the material in question. However, as suggested by the ballistic examples, they can have a very substantial effect. Adiabatic shear bands commonly act as sites for further damage and may act in either a ductile

Figure 1.16. Adiabatic shear band formed in a compression specimen of Ta at a temperature of 77 K and a strain rate of $5.5 \times 10^3 \, \mathrm{s}^{-1}$. Maximum stress occurred at a strain of <0.05, and localization occurred at ~0.23. Material did not localize at room temperature with the same strain rate although maximum stress occurred at the same low strain. (Chen and Meyers, personal communication of unpublished data; collateral to work published by Meyers et al. 1995.)

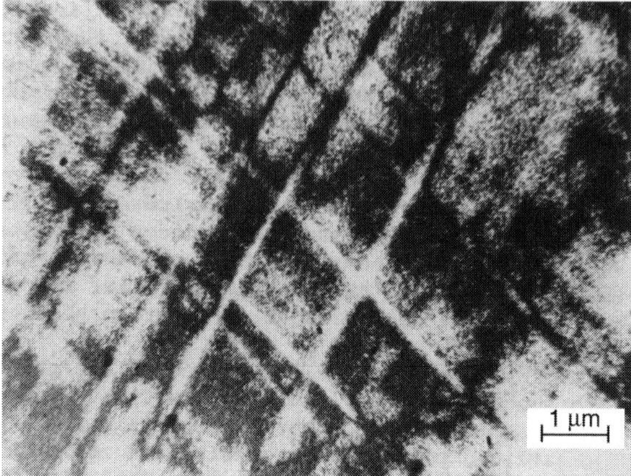

Figure 1.17. Fine scale deformation pattern shown in a Transmission electrons microscopy (TEM) metallograph of Al shocked to 9.0 GPa. (Reproduced from Grady et al. 1983, with permission from Kluwer Academic/Plenum Publishers.)

or brittle manner. Voids sometimes appear within the band, as shown in Figure 1.18, giving the appearance of having experienced tensile stresses while still at an elevated temperature. In contrast, short cracks sometimes also appear within a band (e.g., see Rogers 1983), suggesting that the critical stress had been experienced somewhat later when the material within the band had cooled

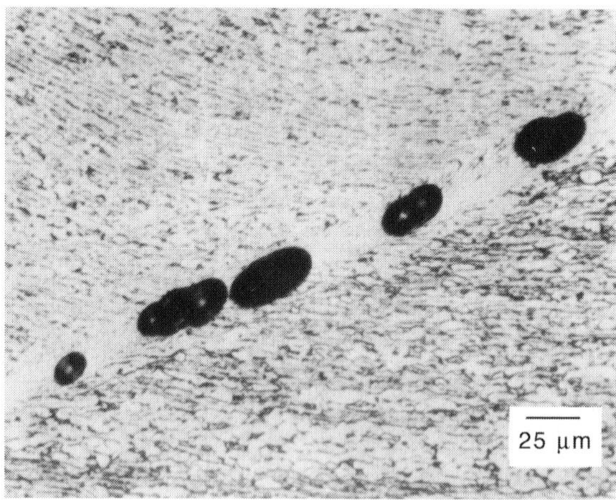

Figure 1.18. Voids formed within a shear band in Ti-6Al-4V. (Reproduced from Grebe et al. 1985. Copyright 1985 by the Minerals, Metals and Materials Society and ASM International.)

and achieved a hardened and more brittle state. In both cases, it is apparent that the locations of shear bands are also sites for possible future failures.

For applications in which a device is to be used repeatedly and must experience many service cycles, it is important that damage sites not be introduced either during its designed use or during its manufacture. For these cases it would be highly desirable to understand the conditions that produce adiabatic shear bands so that they can be avoided. This is generally not difficult for the service lifetime of a device in which plastic deformation itself is to be avoided. However, in modern high-rate manufacturing (forging, impact or electromagnetic forming, etc.), avoidance of the conditions that can produce adiabatic shear bands may be more problematic. Some knowledge of the mechanics of shear bands should allow the rational design of processes that avoid introduction of damage sites. In fact, guidelines for calculating material susceptibility to adiabatic shear have been reviewed (Semiatin et al. 1983 or Semiatin and Jonas 1984).

In contrast, in some industrial processes, such as drilling, cutting, shearing, or punching, as well as milling and machining, material failure by shear is an integral part of the process itself and occurs only once at a given location in the material. It would seem that optimum design of these processes could be more readily achieved with a thorough knowledge of shear mechanics. All the industrial processes mentioned so far have been developed largely by semiempirical means and of course are already in widespread use, but because of very high volumes worldwide, even small added efficiencies could translate into large cost savings.

Nevertheless, it is in ballistics and impact physics that the most immediate and dramatic payoff may be expected from the intelligent use of shear mechanics. In these cases not only is material flow and failure fundamental to the functioning of the service part, but if impact ceases without perforation, these processes must have been operating just beyond a limiting condition. Furthermore, as was suggested by the example of penetration, the extent and distribution of damage may be important in optimizing performance, as well as in minimizing the weight and expenditure of energy that are required for effective functioning. True optimization will not be achieved in the complex devices under consideration today without a full understanding of damage mechanics, of which adiabatic shear is a principal and often dominant example. Even design for crashworthiness of automobiles, aircraft, ships, and other vehicles (which generally requires consideration of lower levels of stresses and strain rates than those attained in ballistics) may well benefit from greater understanding of failure by adiabatic shear.

1.1.4 Qualitative Mechanics

In a qualitative sense, the basic mechanics of adiabatic shear is easily understood. As plastic deformation develops, the isothermal flow stress generally increases with work hardening in most metals, and strain rate hardening may increase the flow stress still further. However, plastic work is converted mostly into heat, and as the temperature in the material increases, the flow stress usually tends to decrease. Thus, competing mechanisms are at work: work hardening and rate hardening tend to raise the flow stress, and thermal softening tends to lower the flow stress. Thermal softening almost invariably wins out over all other hardening mechanisms, so that if deformation continues long enough, eventually the material softens with increasing strain. The competing isothermal mechanisms and a typical adiabatic path are sketched in Figure 1.19.

Strain softening, if strong enough, becomes unstable so that small disturbances in the flow can accelerate, increasing the plastic work and heat generation locally, softening the material still further, and finally drastically altering the ability of the material to transmit shear forces. As the instability develops, a "postbuckling" structure forms, and the fully formed shear band takes shape. This process is generally referred to as localization. The reduced stress in the band now causes unloading in adjacent material where shearing virtually

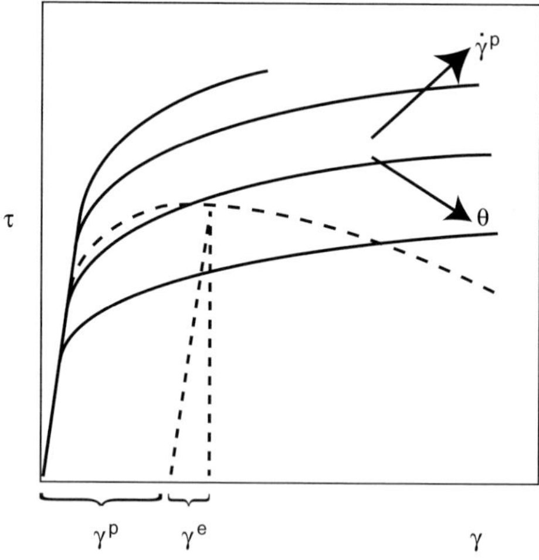

Figure 1.19. Schematic stress–strain curves showing that the isothermal curves (solid lines) tend to lie at higher levels with increasing strain rate, and at lower levels with increasing temperature. The dashed line shows a typical curve for adiabatic loading. The curve starts along an isothermal path at a constant strain rate, but as plastic work and heating build up, the stress reaches a maximum and strain softening sets in.

ceases. In the final state all shearing is confined essentially to the band itself, which becomes very hot; the materials on either side simply translate relative to each other as rigid bodies, and heat generation from the intense, local shearing within the band is balanced by heat conduction to the adjacent, cooler material. In this final configuration, the deformation is anything but adiabatic, so the phrase "adiabatic shear band" is actually a historical misnomer. Much of the early deformation can often be regarded as adiabatic because high strain rates often do make heat conduction unimportant at that stage, but once the instability and localization set in, the temperature gradients to the sides of the band become extremely large, and the deformation can no longer be regarded as adiabatic. If the driving forces that caused the band to form now cease, shearing and heat generation within the band must also stop, and heat conduction causes a rapid drop in peak temperature within the band.

The qualitative description above accounts in a rough way for many of the dominant features seen in Figures 1.1–1.18. Instability and localization account for the thinness of shear bands. Stress drop and unloading of adjacent material account for the "frozen-in" strain patterns. High strains and work hardening accompanied by rapid heating and cooling account for the high hardness often seen within a shear band, although the pattern may sometimes be altered by the exact cycle of straining and heating, even including recovery and recrystallization. Voids or cracks can form within a shear band depending on the state of the material when loads additional to the basic shearing are applied, with tension promoting voids in a hot state and cracks if sufficient cooling, hardness, and brittleness have developed. Finally it is again apparent that adiabatic shear bands can form in many different materials and in many different configurations, not just from stress concentrations, if the loading and heating cycles are favorable.

1.1.5 Reviews and Symposiums on Adiabatic Shear

The overview developed here is a distillation of the work of many researchers over a considerable length of time. In the nineteenth century, perhaps Tresca (1878) was the first to comment on the appearance of an adiabatic shear band. He described the formation of a thermal cross in forging an alloy of Pt-Ir, as has been pointed out by Bell (1974) and Johnson (1987). Tresca recognized that strength ("requires for its deformation a large quantity of work"), low thermal conductivity, and low heat capacity all contribute to the phenomenon. Others have also observed thermal crosses in forging (see references in Johnson 1987). However, it is Zener and Hollomon (1944) who are usually credited with recognizing the distinctive character of adiabatic shear bands as the residual signature of a local material instability, and that the instability is powered mainly by the dominance of thermal softening as a response to heating from plastic work.

In the next three decades or so following the important observations of Zener and Hollomon, progress in the study of adiabatic shear bands was made more from the perspective of materials science than from the point of view of mechanics. Useful reviews have been given by Rogers (1979, 1983), Rogers and Shastry (1982), Dormeval (1988), and Zurek and Meyers (1996). One early attempt to study the mechanics of adiabatic shear was the work of Recht (1964). Another major effort from Erlich, Seaman, and Shockey (1980) tried to blend materials and mechanics by postulating a nucleation and growth mechanism for shear bands, which was calibrated by extensive metallographic observations (also see Curran, Seaman, and Shockey 1987). Although they did not come to grips either with the basic mechanics of individual band formation or with the properly invariant form of a continuum damage theory, they were the first to construct a large-scale code that included shear damage.

In 1980 the U.S. National Materials Advisory Board published a study, headed by W. Herrmann, which focused on the phenomena of high rate deformation (National Materials Advisory Board 1980). Included in that study was a section on adiabatic shear bands (Clifton 1980), which inaugurated a great increase of activity in the mechanics of adiabatic shear. Since 1980 too large a volume of theoretical and experimental work has appeared in the engineering literature to be adequately summarized in this brief section, except to call attention to several of the principal reviews and symposiums on the topic. The Sagamore Conference of 1982 dealt with recent advances in dynamic material behavior and included significant work on adiabatic shear (Mescall and Weiss 1983). A review by Shawki and Clifton (1989) covered analytical work up to that time and included some computations for a thin-walled tube with a flat notch to trigger localization. Dormeval (1988) reviewed metallurgical aspects and observed that many dynamical measurements will be required to verify theoretical and numerical results. Appropriate experimentation and verification still have not been done to any significant degree. The most comprehensive summary to date is from Bai and Dodd (1992), whose book on adiabatic shear touches on all aspects of the subject. Meyers (1994) in his book also devotes a chapter to adiabatic shear bands and another chapter to plastic deformation at high rates. Tomita (1994) and Batra (1998) covered computational aspects of the subject. Several recent symposiums, devoted entirely to the topic, were published almost in their entirety and gave snapshots of the activity and accomplishments at the date of publication (Zbib, Shawki, and Batra 1992; Armstrong, Batra, Meyers, and Wright 1994; and Batra, Rajapakse, and Zbib 1994).

In spite of all the activity cited here and much more besides, relatively little analytical work has attempted to treat the fine-scale physics of band formation in sufficient detail that scaling laws could be formulated; that is to say, to make precise statements of the effects to be expected from physical and kinematic parameters. In fact, it is one of the principal aims of this book to fill some of that gap.

1.1.6 Implications for Computations

Because the morphology of an adiabatic shear band exhibits such a fine transverse scale, it would be costly to resolve the bands fully in a large-scale computation. Scaling laws can actually estimate the mechanical resolution required for a given material and loading (see Chapter 7, Section 7.4), and the estimated scales are generally comparable with those seen directly in Figure 1.3. Both estimated and observed scales indicate that a resolution of the order of 10 μm or less would be required for many materials in order to follow the deformation across the band in detail. Furthermore, if thermal conductivity is not included, at least in regions very close to a shear band, mesh sensitivity will always occur, and no fully converged, numerical solution will be found at any level of mesh refinement. This point will be made clearer in Chapter 7.

In many published calculations, not only is thermal conductivity neglected, but the mesh is as much as an order of magnitude larger than the scale to be expected within the fully formed shear band. This means that the computed localization will have essentially the same width as the grid resolution at the band, and it is the length scale of the grid that regularizes the calculation rather than the physical and constitutive features of the material. Probably the major computed shear bands still form near the locations where they would actually occur, but the accuracy of timing, propagation, transmitted forces across bands, and influence on other regions must all be called into question. Other calculations with heat conduction included have fully resolved the bands, but only in one spatial dimension and only in a relatively narrow strip that includes the band. Some of these calculations, which may be classified as research studies on the dynamics of band formation, are discussed more fully in Chapter 5, Section 5.4.

Because the formation of adiabatic shear bands has a major effect on failure processes, and because the scale at which these bands form is far below anything that is convenient or economic to resolve in production calculations, the designer who needs to include dynamic failure as an integral part of the design is faced with a major dilemma. It is actually a common situation in continuum physics that important physical processes occur at an inconveniently fine scale. Besides failure processes, other examples are turbulence and boundary layers in fluids, droplet formation and combustion in mixtures of liquid fuel and air, hot spot formation in explosives, or storm cells in weather prediction. One resolution of the dilemma is to find a way to get the effect of the important local process without doing the work of full resolution each and every time that local effects are called for. Usually the solution can be found by careful study of the local process, extracting essential features in some simplified manner, and then inventing a method to inject the results into the larger-scale problem in a realistic way. A phrase often used today to describe this procedure is "bridging the length scales." An elaborate example, in which several length scales were involved in designing a high-performance steel with both improved strength and toughness, is described by Olson (1997).

This volume is intended to contribute to the detailed study of the local processes in shear band formation and to extract many of the major features through scaling laws. At present, however, the method of injecting these results into a larger calculation so as to complete the bridge is still in its infancy. Two approaches appear to be practicable. In one, a shear band is treated as a boundary layer between two parts of the material, and rules are given for the interaction of the two parts across the boundary. Walter and Kingman (1998), adapting a method from Silling (1992), showed an example in which the presence of a shear band permitted a high-speed particle to split in two upon oblique impact with a thin plate rather than flattening out as a result of gross plastic flow. The method is still somewhat artificial in that it requires arbitrary seeding and propagation of bands instead of being triggered and driven by critical physical conditions. However, it does show that the presence of a shear band can have a major effect on the results of the calculation. In the second approach, creating "damage" in the continuum response simulates the effect of shear band formation. Raftenberg (2000) used an initiation criterion to trigger isotropic shear damage, which he simulated by artificially reducing the flow stress and spall strength to arbitrary lower levels. This had a strong effect on the results of his calculations, but the model requires calibration of several parameters, and it appears to be difficult to tie to the actual physical process. The method of Erlich et al. (1980), mentioned previously, also uses the concept of volumetric damage, but additionally it retains the directionality of shear damage in a discrete way.

It appears that an approach using either boundary layers or volumetric damage can probably be extremely fruitful if adequately coupled with the physics of the local processes.

1.2　One-Dimensional Experiments

The basic hypothesis concerning the formation of adiabatic shear bands, as stated in the last section, is that they occur after rapid plastic flow has become unstable and that their extreme postlocalization configuration results by attaining new dynamic balances of force and energy. In order to turn the qualitative description of the last section into a truly quantitative understanding, experiments must be designed and carried out both to establish the basic properties of the material and to exhibit the formation of adiabatic shear bands as a consequence of those properties. Furthermore, the experiments must not only be relatively easy to perform, but they must be susceptible to relevant measurements and quantitative interpretation. Because it often happens that ease of performing an experiment and ease of quantitative interpretation are inversely related to each other, experimental design invariably involves compromise and is something of an art. For example, although it is easy to form adiabatic shear bands in impact experiments, little information that is both useful and quantitative can be reliably established during the time of formation. Postmortem examination does supply useful information, of course, but it does not supply much illumination concerning the dynamics of formation.

Material testing to determine constitutive response usually depends implicitly upon the assumption of local action. That is to say, the stress at a material particle is assumed to depend only on the values of field variables and perhaps their histories at that same particle. Then if a volume of material can be placed in a uniform state, point values and average values over that volume will coincide so that measurement of the average values of all relevant fields will suffice to establish the desired pointwise relationships. The field of dynamic measurement of this kind has been advancing rapidly in recent years. Other constitutive behaviors, involving gradients or nonlocal action for instance, have also been assumed, but they are more difficult to establish experimentally. If the phenomenon to be studied is inherently dynamic and nonuniform, as is the case with adiabatic shear bands, then the difficulties are compounded, and methods must be devised so that individual bands may be studied in relative isolation. Nevertheless, considerable progress in this area has been achieved as well.

1.2.1 Dynamic Testing With Uniform Fields

The standard technique for some time has involved use of the split Hopkinson bar (or Kolsky bar), which today is used in three basic variations: compression, torsion, and tension. The theory and practice in use of this device has been described by Lindholm (1964) and more recently and completely by Nicholas (1982), Nicholas and Bless (1985), Follansbee (1985), Hartley, Duffy, and Hawley (1985), and Field, Walley, Bourne, and Huntley (1994), as well as others. The basic idea is simple and is the same for all three variations, although practical considerations vary greatly among the three and can be quite intricate. (See the references for details.) A cylindrical specimen (solid cylinder for tension and compression or thin-walled tube for torsion) is placed in line with and between the ends of two high-strength elastic bars and in contact with both, as sketched in Figure 1.20. Then a trapezoidally shaped stress pulse of

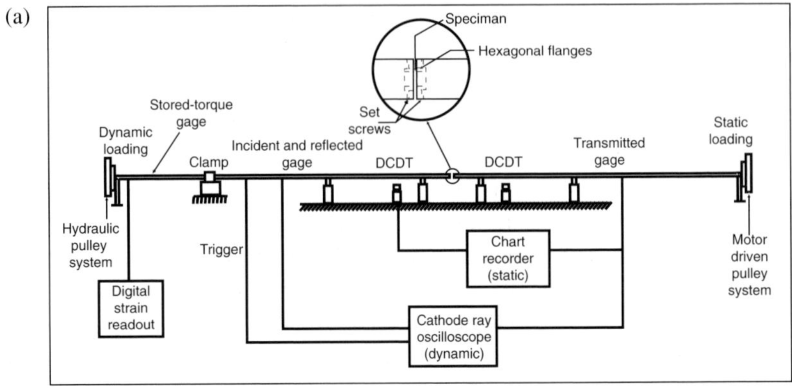

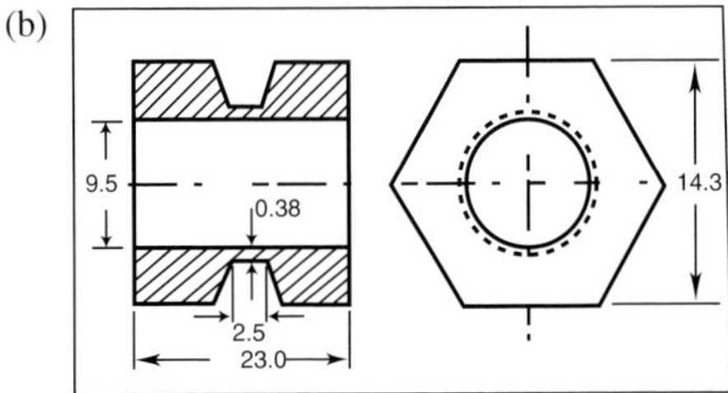

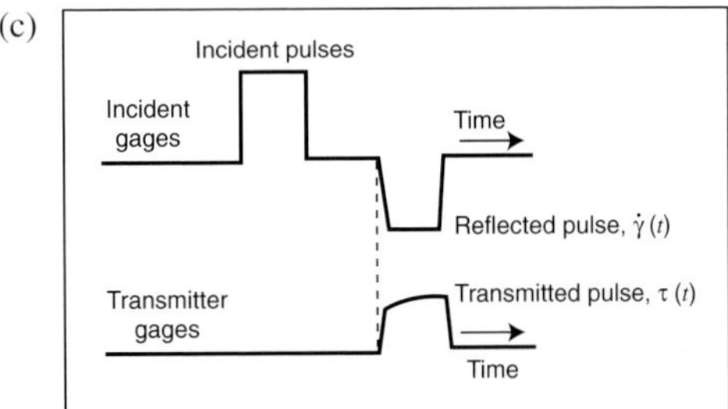

Figure 1.20. Panel a shows a schematic diagram of a Kolsky torsional bar apparatus. Panel b shows a typical thin-walled torsion specimen. Panel c shows typical ideal waveforms to be expected at the incident and transmission gages. (All three panels are reproduced from the *ASM Handbook, Mechanical Testing*, in the article by Hartley, Duffy, and Hawley 1985, with permission from ASM International.)

controlled shape, length, and intensity is made to propagate down the input bar toward the specimen and the output or transmission bar. The specimen must be short enough and ductile enough that it can come into equilibrium with the forces in the two bars during the early part of the loading pulse. It must also have a smaller cross-sectional area and smaller bar impedance so that it can be driven into plastic deformation while the two bars remain elastic. Then by making separate measurements of the loading, reflected, and transmitted pulses in the elastic bars, and by using the linear theory of one-dimensional elastic wave propagation in a bar, one can deduce the force transmitted through the specimen and the relative velocity between the two ends of the specimen for as long as the loading pulse lasts. Because the initial geometry of the specimen is known, stress and strain rate in the specimen are then easily calculated, and strain follows by integration. Refinements for finite deformation in tension or compression have been developed by Ramesh and Narasimhan (1996).

Data are usually presented as a dynamic stress–strain curve for the specimen at a nearly constant strain rate and temperature (e.g., see Figure 1.21). The oscillations in the early part of the curve indicate undesirable effects from wave propagation and usually tend to become more severe as the applied strain rate increases. Actual strain rates may vary by ±20% from the nominal rates during a test, and the temperature may increase by several tens of degrees Celsius at large strains. For a given bar and specimen the imposed strain rate in the specimen can be varied by changing the intensity of the loading pulse, and for a given strain rate the final strain can be varied by changing the length of the pulse. Strain

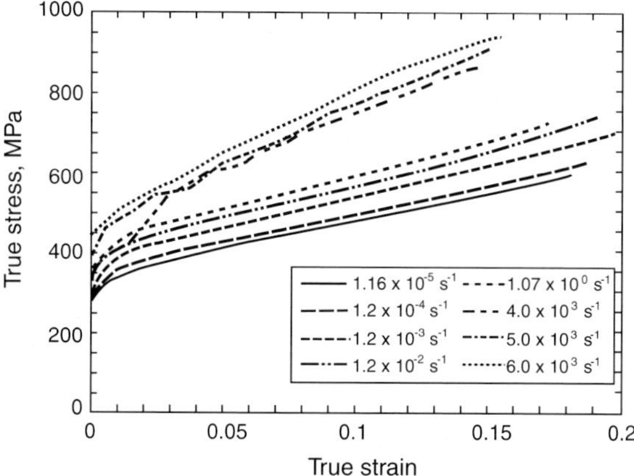

Figure 1.21. Quasi-static, low strain rate, and high strain rate data for alpha titanium. (Reproduced from Chichili, Ramesh, and Hemker 1998, with permission from Elsevier Science.)

rate can also be changed by varying the length of the specimen because smaller size means higher rates, all else being the same. A large change in length of the specimen, however, usually also requires a redesign of the whole apparatus because scales and dimensions have to be properly balanced to achieve the best results. Changing the initial temperature of the specimen is also possible, but this requires special techniques as noted below.

Much of the difficulty of use and the setting of operational limits in a split Hopkinson bar have to do with the maintenance of nearly uniform conditions over a usable effective gage length within the specimen. Some considerations have to do with overall design of the experiment and some have to do with secondary, but potentially very important, dynamical effects.

As an example of the first type of consideration, all test specimens must be short enough that they can come into equilibrium in the early part of the deformation, as mentioned in the opening paragraph of this section. Because shear stress varies in torsion with the radius in a solid bar or thick-walled tube, thin-walled tubes are used so that the stress will be nearly uniform. In torsion or tension tests there must be an adequate ratio of length to thickness in the nominal gage section of the specimen so that transitional end effects still permit a usable effective gage section. As another example, high-quality lubrication is used on the ends of compression specimens so that friction may be reduced as much as possible to eliminate barreling. Other considerations may be found in the references.

Even if uniform conditions have been established in the axial direction, secondary dynamical effects may cause nonuniformity in other stress components. For example, in tension and compression specimens the radius of the specimen limits the usable axial strain rate because the imposed axial rate induces a nonuniform radial acceleration and hence a nonuniform radial stress in the specimen. The maximum of this induced stress must be kept to a small fraction of the measured axial stress if the latter is to be interpreted as the flow stress. On dimensional grounds or from a simple dynamical analysis it can be argued that the required condition is $a\dot{\varepsilon} \ll \sqrt{Y/\rho}$. In words, the product of the axial strain rate and the radius of the specimen must be much less than the square root of the flow stress in the specimen divided by its density. (The symbol $\ll$, meaning "much less than," usually can be safely interpreted to mean "less by at least a factor of 5.") In torsion tests on a thin-walled tube, centrifugal forces will induce hoop stresses in the center of the gage section and radial shear stresses at the ends. To keep these far below the flow stress in shear, simple analysis again suggests that $h\dot{\gamma} \ll \sqrt{Y/\rho}$ for hoop stresses and $h^3\dot{\gamma}^2/a \ll Y/\rho$ for shear stresses. In these two expressions h is the gage length of the thin tube, $\dot{\gamma}$ is the shear strain rate, and all other terms are the same as before. More exact conditions must be determined by computation, iterated with experimentation.

For typical designs and materials the applied strain rate usually must be kept well below a theoretical value of 10,000 s^{-1}, although a miniature design of a direct Hopkinson bar has permitted higher rates in compression (Gorham, Pope, and Field 1992). Others have developed miniature split Hopkinson bars, as well. Practical considerations, such as the ability to apply a frictional grip to a torsion specimen so as to store high torque, may also limit the achievable strain rate to a level well below the theoretical limit. Barreling in compression, necking in tension, and buckling of the wall in torsion all signify that uniform conditions of stress and strain have not been maintained in the effective gage section. When those conditions occur, the data can no longer be interpreted as the fundamental constitutive response of the material, although the data can still be interpreted as force transmitted and relative velocity across the specimen.

A review paper by Duffy (1980) discusses typical uses and data obtained. The major use of the split Hopkinson bar in all its forms has been the observation of variations in the stress–strain response for a material as the applied strain rate and temperature are varied. When stress is cross-plotted against strain rate at some fixed strain (say 0.05 or 0.10) and temperature in a log–log plot, the slope of the curve is referred to as the strain rate sensitivity. An example is shown in Figure 1.22. Typical values for structural metal alloys at room temperature are less than 0.03 and nearly constant over a decade or more of strain rate. Strain rate sensitivity tends to increase with temperature for a given material, and a pure metal generally has a much higher sensitivity than its structural alloys. The simplest interpretation of data from systematic tests with a split Hopkinson bar,

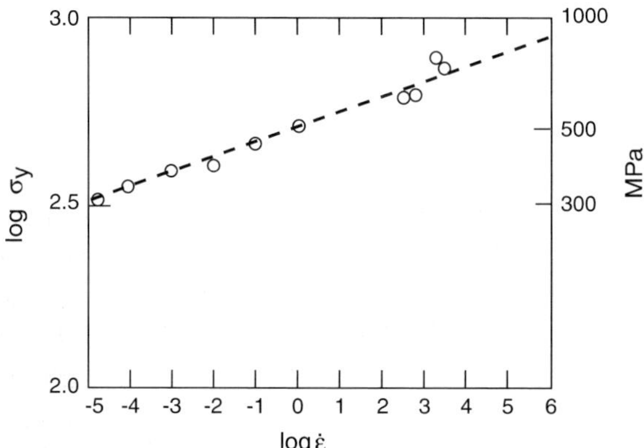

Figure 1.22. Strain rate sensitivity is determined from the slope of a log–log plot of stress vs. strain rate at constant strain and temperature. Data here are for a stainless steel and seem to show a tendency toward increased sensitivity between 10^3 and 10^4 s^{-1}. (Reproduced from Kassner and Breithaupt 1984, with permission from IOP Publishing Limited.)

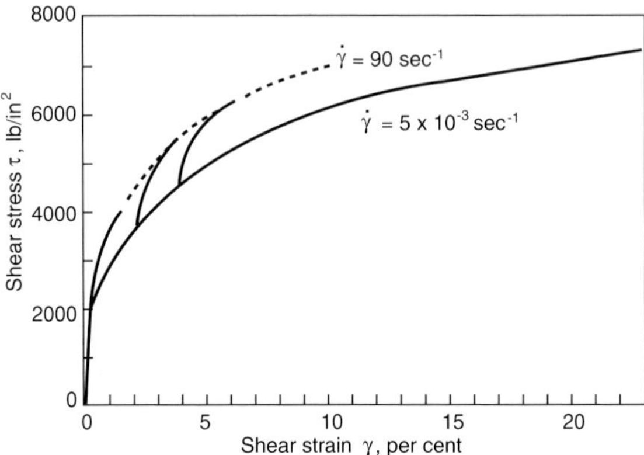

Figure 1.23. Incremental stress–strain curves (jump tests) for torsional specimens of Al. After a short transient the jump response is asymptotic to the higher rate curve, showing no "history" effect, but note that the higher rate is only 90 s^{-1}. (Reproduced from Campbell and Dowling 1970, with permission from Elsevier Science.)

and the one most commonly invoked, is that the flow stress may be represented as a function of plastic strain and plastic strain rate, as well as temperature. Such an interpretation is only a simple extension of the quasistatic stress–strain curves of long familiarity.

Other tests have hinted that the situation might be more complex, at least for some materials. If distinct stress–strain curves for a given material are obtained at two different strain rates, then the simple interpretation of the last paragraph would predict that a third experiment that switched from one strain rate to the other in the middle of the test would simply switch from one curve to the other after a short transient, as shown in Figure 1.23. However, in so-called jump tests, also described in Duffy's review article, it has been observed that the simple behavior does not always occur. The jump test starts out on one of the previously determined curves, but when the rate is changed the response sometimes either overshoots or undershoots the other known curve, as shown in Figure 1.24. Such behavior has been labeled a "history effect." Chapter 4, Section 4.1 suggests that it may be modeled with the aid of evolving internal variables and so may not actually require the history of any independent constitutive variable, but only current values. However, the experimental situation has to be critically reviewed and more data taken before definitive theoretical improvements can be made.

As theory and computations advanced during the past two decades, experimental work advanced, too. Interest extended to more extreme conditions of temperature, strain, and strain rate, and these extensions required new techniques and refinements of older techniques. Correction for dispersion in axial loading pulses was introduced by Follansbee and Frantz (1983) and has now

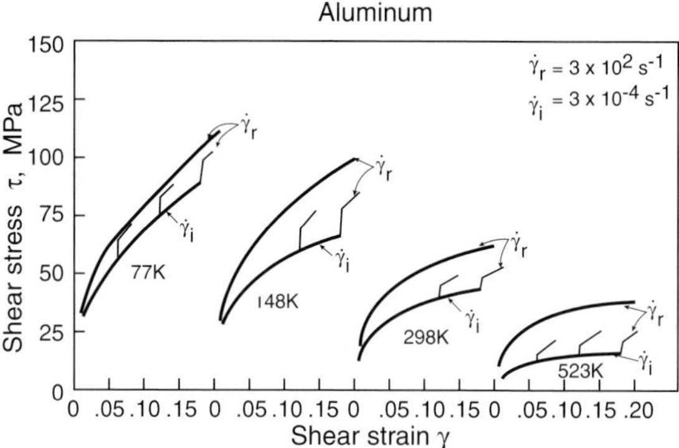

Figure 1.24. Incremental stress–strain curves (jump tests) for torsional specimens of 1100-O aluminum. These tests show a "history" effect that appears to be stronger at higher temperatures and larger initial strains for the jump. (Reproduced from Senseny, Duffy, and Hawley 1978, with permission from ASME.)

become standard procedure in many laboratories (e.g., see Coates and Ramesh 1991). In order to minimize mass diffusion and other slower metallurgical changes that might cause additional changes in the constitutive response, rapid heating techniques have been developed. Gilat and Wu (1994) used direct flame, Lennon and Ramesh (1998) used infrared radiation, and Nemat-Nasser and Isaacs (1997) used a small induction furnace to achieve high temperatures quickly. A feature of the last two works is that the loading bars are brought into contact with the specimen nearly simultaneously with the loading pulse so as to minimize the effects of heat conduction and change of wave speed in the loading bars. As longer pulses are used at high strain rates, the adiabatic response of the material becomes evident. So as to capture both the adiabatic and the isothermal response, Nemat-Nasser and Isaacs (1997) used sequences of interrupted tests. Figure 1.25 shows typical data.

Reflections of the loading pulse from the extreme ends of the apparatus can cause repeated, and unwanted, loading of the specimen so that the full history of loading becomes uncontrollable and poorly known. So as to maintain suitability for postmortem examination, Nemat-Nasser, Isaacs, and Starrett (1991), among others for example, devised means of trapping and eliminating the pulse after one pass through the specimen in compression and tension tests. Thus, the loading becomes well known and well characterized for correlation with subsequent metallurgical study.

Testing at the extreme ranges for split Hopkinson bars gave indication that strain rate sensitivity might increase substantially at higher rates (see, e.g., Follansbee, Regazzoni, and Kocks 1984), although secondary dynamical effects cannot be excluded. With the introduction of the transverse displacement

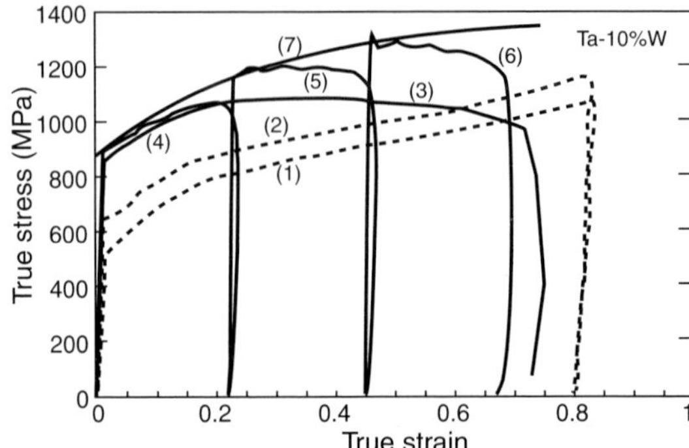

Figure 1.25. Stress–strain curves for Ta-10W. Curves (1) and (2) are isothermal curves at a temperature of 25 °C and strain rates of 10^{-3} and $1\,\mathrm{s}^{-1}$, respectively. Curve (3) is an adiabatic curve at $5700\,\mathrm{s}^{-1}$ beginning at 25 °C. Curves (4), (5), and (6) are also adiabatic curves at $5700\,\mathrm{s}^{-1}$ beginning at 25 °C, but interrupted and resumed at the strains shown after returning to 25 °C. Curve (7) is the estimated isothermal curve at $5700\,\mathrm{s}^{-1}$ and 25 °C. (Reproduced from Nemat-Nasser and Isaacs 1997, with permission from Elsevier Science.)

interferometer (Kim, Clifton, and Kumar 1977), it became possible to conduct pressure–shear tests in which a material is subjected to shock loading in a gas gun with both normal and transverse components of velocity. Later, by sandwiching a thin specimen between two hard, elastic materials, it became possible to conduct high strain rate tests in shear at rates of $10^5\,\mathrm{s}^{-1}$ and up, more than an order of magnitude higher than is possible in the standard Hopkinson bar test. A review of this method and some of the data to that time were given by Klopp, Clifton, and Shawki (1985). Figure 1.26 shows schematically a sketch of a typical experimental setup. Strain rate sensitivities were found to be much

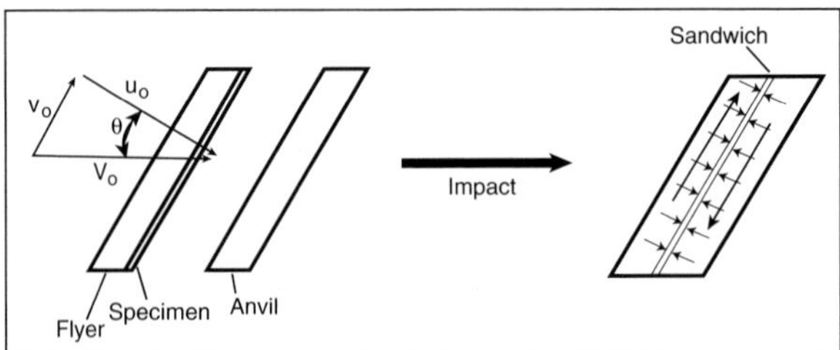

Figure 1.26. Schematic diagram of pressure–shear experiment. After impact, the arrangement of a plastically deforming specimen between hard elastic materials is very similar in concept to the Kolsky bar. The thinness of the specimen allows shear strain rates that are typically more than an order of magnitude higher than in a torsional Kolsky bar, however. (Reproduced from the *ASM Handbook, Mechanical Testing*, in the article by Clifton and Klopp 1985, with permission from ASM International.)

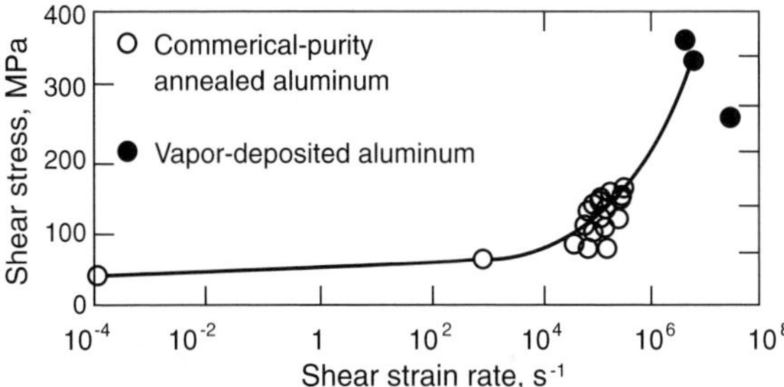

Figure 1.27. Strain rate sensitivity for Al appears to increase dramatically at the higher strain rates achievable in pressure–shear tests. Similar results also hold for other materials. (Reproduced from the *ASM Handbook, Mechanical Testing*, in the article by Clifton and Klopp 1985, with permission from ASM International.)

higher under these conditions (Figure 1.27), thus apparently confirming the earlier work, although differences in microstructure and material condition cause some confusion in the literature. Overlapping data from Hopkinson bar tests and pressure–shear tests is lacking, however, and there is no means as yet for testing at intermediate rates that also overlap the other two methods. Means to conduct dynamic tests with combined compression and torsion have also been devised for a Hopkinson bar configuration by Chichili and Ramesh (1999). These experiments also incorporated a scheme to eliminate wave reflections and repeated loading of the specimen.

1.2.2 Formation of Adiabatic Shear Bands in Thin-Walled Tubes and Other Geometries

As shear strains and strain rates are increased in torsional Hopkinson bar testing of thin-walled tubes, conditions become favorable for the formation of adiabatic shear bands as an instability that develops in adiabatic plastic flow. The force transmitted through the specimen may still be interpreted as stress if the gage length of the specimen is short enough and the rate of loading is slow enough that quasi-static conditions prevail, but as localization develops, the velocity difference across the gage length can only be interpreted as a nominal strain rate. The temperature in the specimen also becomes highly nonuniform as the material in the shear band heats up by hundreds of degrees Celsius.

Costin, Crisman, Hawley, and Duffy (1979) published extensive data for two steels, tested in torsion at strain rates of $500\,\mathrm{s}^{-1}$ and $1000\,\mathrm{s}^{-1}$ and at temperatures from $-157\,^{\circ}\mathrm{C}$ to $+121\,^{\circ}\mathrm{C}$. The hot-rolled 1020 steel showed extensive work hardening but no localization, whereas the cold-rolled 1018 steel showed little work hardening and tended to localize at the higher strain rates. In these experiments the distribution of strain was measured only after the test by examining an axial line that was scribed on the inner surface before the test. Marchand and Duffy (1988) added a nonuniform grid along the exterior generators of the gage section so that displacements could be observed directly during localization. They also measured the temperature profile across the localized zone by an infrared technique. As a result of these more complete measurements, they were able to describe the process in great detail, as shown in Figure 1.28.

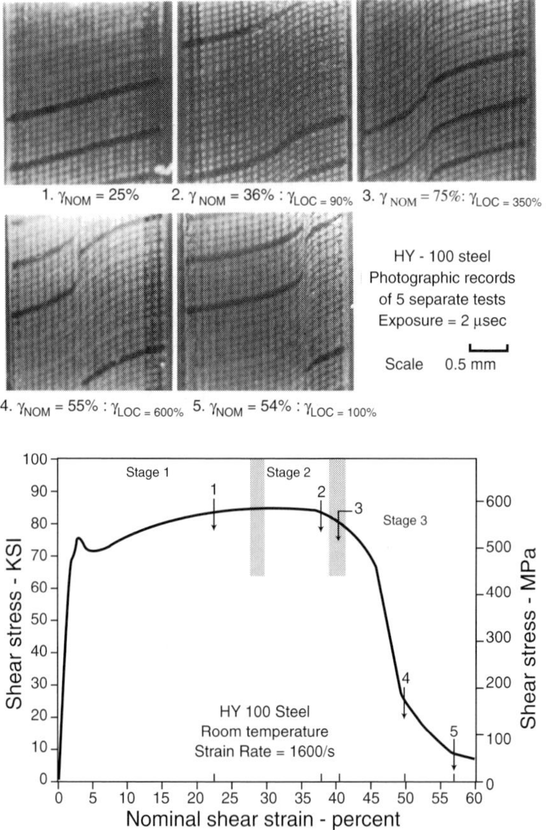

Figure 1.28. Stress–strain curve in torsion of HY 100 steel showing localization that accompanies formation of an adiabatic shear band. In stage 1 the strain is uniform throughout the shearing section; in stage 2 nonuniformity begins and increases as the stress passes through a maximum; and in stage 3 the stress drops sharply as the temperature and strain in the localized region increase dramatically. The photos above the stress–strain curve show the distortion of a grid laid down on the side of the thin-walled specimen. Photos were taken in separate tests but correspond roughly to the arrows shown on the curve below. (Reproduced from Marchand and Duffy 1988, with permission from Elsevier Science.)

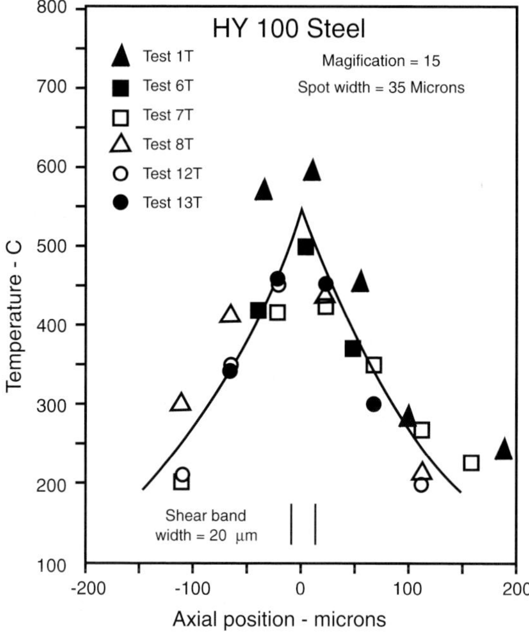

Figure 1.29. Temperature profile measured across an adiabatic shear band at a late stage of deformation. (Reproduced from Marchand and Duffy 1988, with permission from Elsevier Science.)

Nearly uniform deformation develops at the beginning as the flow stress increases with initial work hardening. During this stage, temperature increase is moderate. Next the flow stress, showing little variation as strain accumulates, passes through a weak maximum as deformation begins to show some nonuniformity, and temperature continues its moderate rate of increase. Finally the stress drops rapidly and the temperature in the center of the band increases dramatically, as the deformation becomes so nonuniform that it is difficult to observe the structure of the band, except for the relative displacement from one side to the other. The temperature profile across the band is shown in Figure 1.29. Although the resolution of the instrumentation was not fine enough to make precise observations, there is a strong impression that the thermal profile across the band at a late stage is sharply cusped, as indicated in the figure. Shortly after the stress drop or even during the drop, the specimen usually breaks.

At about the same time Giovanola (1988) also made more highly resolved measurements in a Hopkinson bar test and came to similar conclusions; see Figure 1.30. Progress in theoretical studies resulted in a similar description, as well (Molinari and Clifton 1987, Wright and Walter 1987, and Wright 1990a).

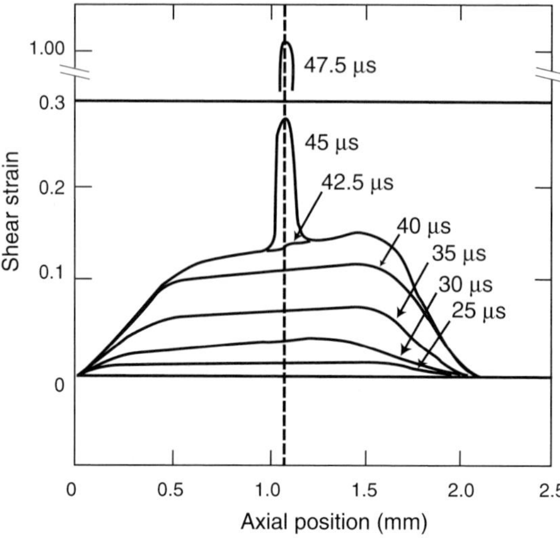

Figure 1.30. Strain profiles across an adiabatic shear band in 4340 steel (HRC 40). Displacements were measured directly by photographing a fine grid laid down on the side of the torsion specimen. Strains were calculated from the photographs. (Reproduced from Giovanola 1988, with permission from Elsevier Science.)

Through the 1980s most workers assumed that instability of adiabatic flow alone was sufficient to explain the formation of adiabatic shear bands, although in a perturbation analysis Anand, Kim, and Shawki (1987) included pressure as well as thermal sensitivity. In an experimental study Cowie, Azrin, and Olson (1989) observed that voids also formed along the sheared zone in double shear tests. They used a split Hopkinson bar to shear the center section of a small beam relative to its ends. The geometry of the test undoubtedly introduced some bending and tension transverse to the shear planes. As might be expected, they also found that an imposed normal stress (an axial stress in the beam) delayed localization, presumably by suppressing void nucleation. Later Weerasooriya and Beaulieu (1993) observed void formation in torsion tests of a W metal matrix composite, and Chichili (1997), who also imposed a static axial stress during his tests, observed it in Ti as well.

Void formation generally requires a tensile stress at the point of formation. Metals, when observed on a fine enough scale, contain many nonuniformities of structure, such as misaligned grains, second phases, dispersed particles, and so forth. Therefore, there may be large stress fluctuations and many opportunities for a local tensile stress to develop even though there is no tensile component in the average stress tensor. Furthermore, in the standard torsion test with no axial stress the principal stresses occur at $\pm 45°$ to the plane of maximum

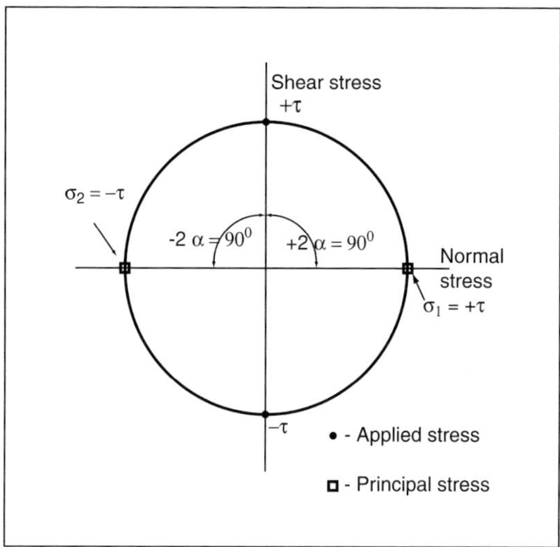

Figure 1.31. Mohr's circle showing that in a thin-walled specimen with a shear stress τ, but no axial stress, the principal stresses are $\pm\tau$ and lie at an angle of $\pm45°$ to the plane of maximum shearing.

shear. One of those principal stresses is tension that is equal in strength to the maximum shear stress, as consideration of Mohr's circle readily shows (Figure 1.31). Weerasooriya (personal communication) has proposed that adiabatic shear, when considered on a diagram of maximum stress and critical strain, somewhat similar to an Ashby map (Ashby 1992), should contain a region of void assisted shear, as well as pure adiabatic shear.

Besides standard testing in a torsional Hopkinson bar, the compression Hopkinson bar has also been used to develop other types of rapid shear test. Hartmann, Kunze, and Meyer (1981) developed the hat test with a configuration of the specimen as shown in Figure 1.32, and Klepaczko (1994) developed a modified double shear test with the configuration shown in Figure 1.33. In either case the test measures the force transmitted through the specimen and the velocity difference between the faces of the load and transmission bars, as expected, but the interpretation of the measurements in terms of shear stress and strain in the specimen is not as straightforward as in the torsional test. Nominal shear stress and relative displacement are readily available, but more detailed determinations of stress and strain are not. Notice that the sheared zones are parallel to the axis of the bar, and that they emanate from a corner or a notch in the specimen. Therefore, the bands originate from a point of stress concentration, and stress will not be uniform along the full length. Furthermore, the bands terminate on the surface so they must

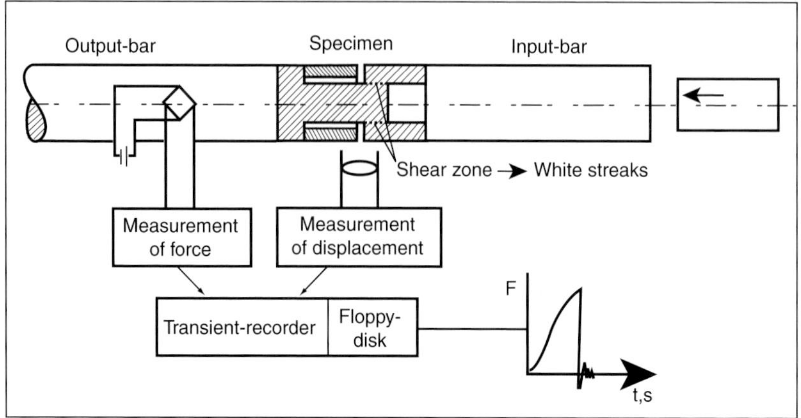

Figure 1.32. Schematic diagram showing an experimental setup for producing adiabatic shear bands with a compression Kolsky bar. A special axisymmetric specimen is designed to shear on a cylindrical surface that is parallel to the axis of the load bar. The extent of deformation can be limited by a stopper collar. Displacement and axial force are measured as shown. (Reproduced from Hartmann et al. 1981, with permission from Kluwer Academic/Plenum Publishers.)

experience end effects. Computations are often needed to interpret the data more precisely.

The torsional Hopkinson bar, which seems so straightforward at first, can also contain hidden complexities. The gage section of the specimen, as used by Duffy and coworkers, was polished to remove possible surface defects from machining, but the act of polishing itself can introduce an unknown defect. Because of the configuration of the specimen, shown in Figure 1.20, it is understandably easy to remove more material from the center of the gage length than from the ends, thus introducing an unknown variation in the wall thickness of the thin-walled tube. Because defects have a strong influence on the timing

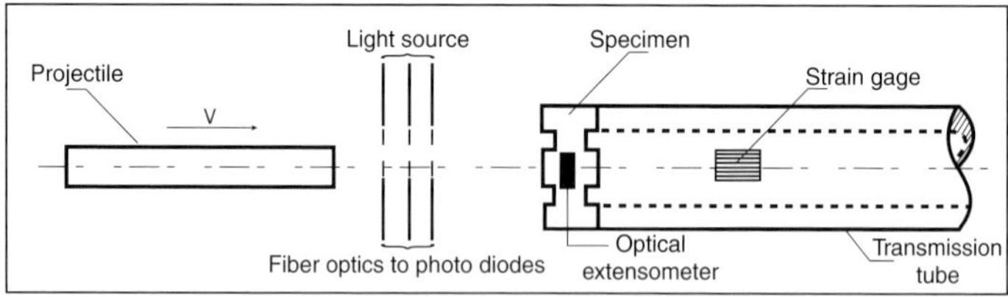

Figure 1.33. Schematic diagram showing an experimental setup for producing adiabatic shear bands with a modified compression Kolsky bar. The specimen is a small beam with two prenotched sections. Direct impact produces rapid shearing of the beam with transmitted forces being measured in the elastic transmission tube. (Reproduced from Klepaczko 1994, with permission from Elsevier Science.)

and strain at which full localization occurs (see Chapters 6 and 7), the timing of stress collapse was being inadvertently manipulated without the knowledge or intent of the experimentalist. Later Duffy and Chi (1992) and Liao and Duffy (1998) demonstrated that the timing of stress collapse could be extended by decreasing the size of a controlled defect.

Lindholm, Nagy, Johnson, and Hoegfeldt (1980) developed an open loop, hydraulic device for torsional testing that used specimens similar in design to those used in the torsional Hopkinson bar, but testing was done at slower rates. Data were collected on a considerable variety of materials, but one intriguing test on Cu at the relatively slow rate of approximately 330 s^{-1} showed alternating localization and work hardening as the stress first began to drop, then partially recovered, then began to drop again, and so forth; see Figure 1.34. In this particular case, the data could not be used to isolate material properties, but they could be interpreted computationally if the material behavior was assumed to be known; see Johnson, Hoegfeldt, Lindholm, and Nagy (1983). Walter (1992) also examined this case numerically and found results similar to those computed by Johnson et al.

Finally, mention should be made of pressure–shear tests reported by Zhou and Clifton (1997) in which high shear and localization were produced in a metal matrix composite of W grains in a W/Ni/Fe matrix. Here the nominal rate of shear was approximately 5×10^5 s^{-1} over a gage length of only 60–80 μm.

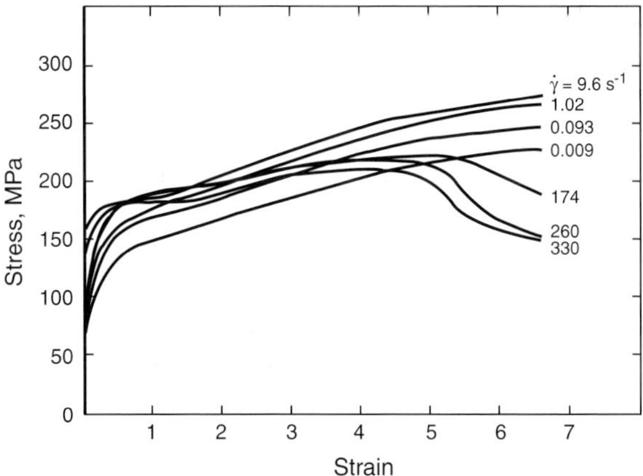

Figure 1.34. Large strain testing of Cu at rates from $\sim$$10^{-2}$ s^{-1} up to $\sim$10 s^{-1} show typical rate hardening of nominally isothermal stress–strain curves. At medium rates of a few hundreds per second, the curves appear to show the beginning of typical adiabatic curves, followed by renewed hardening, and a second softening. (Reproduced from Lindholm et al. 1980, with permission from ASME.)

1.3 Concluding Remarks

In this introduction, adiabatic shear bands have been described in a qualitative way, and the principal experimental means devised to date for their investigation have been briefly reviewed. Perhaps the main point to bear in mind throughout the rest of this volume is that all adiabatic shear bands have a transverse length scale that is extremely small compared with the overall size of the object within which they occur, but that their existence may have a profound effect on the performance of that object. Furthermore, because of their small size and inherent inhomogeneity in temperature, strain, and strain rate, it is all but impossible to understand their full nature or to characterize them completely by experimental means alone. It is here that simple material models and an analytical treatment can be used to help interpret results and even to suggest and guide more profound experimentation. The remainder of this volume is devoted to this task.

2

Balance laws and nonlinear elasticity: A brief summary

One of the most striking features of an adiabatic shear band is the way it cuts through the material as if the microstructure had no influence on its development. This aspect is particularly noticeable in Figure 1.3, where the grain structure and the inhomogeneities in the rolling planes are distorted to conform to the shear band, but the band cuts straight through the structure without deviating from its course. A similar tendency is also noticeable in many of the other photomicrographs in Chapter 1. Only near the tip of a propagating shear band, where it fragments and loses its coherence, does the band seem to be responsive to the details of the local microstructure. On a larger scale, however, the tip of the band merely becomes diffuse and disappears into the general deformation. Consequently, except possibly near the very tip of a shear band, it would seem to be sufficient to regard the band as being controlled by the average thermo-mechanical properties of the material in a larger neighborhood, rather than by the detailed local properties within the individual grains and impurities. This situation is similar to that in conventional fracture mechanics in which the material surrounding the crack is also usually treated as if it were homogeneous and representative of the average properties although a detailed examination of the material at the tip of a crack would reveal otherwise.

The brief discussion in the preceding paragraph is sufficient justification for treating the mechanics of shear band formation in the same fashion as the deformation of the bulk material. Even though the scale of significant variation in strain, temperature, or other fields may only be of the order of 1–10 μm across a shear band, which is often much less than the scale of significant variation in the microstructure of the material, the material will be treated as if it were homogeneous and the usual laws of continuum mechanics and plastic constitutive response applied. Therefore, the remainder of this chapter reviews the conservation or balance laws of mechanics that determine the equations of motion and the laws of thermodynamics that impose restrictions on the material response. These macroscopic laws will then be applied on a scale from a few tenths of a micrometer to a few tens or possibly hundreds of micrometers. This is a small scale, smaller by five or six orders of magnitude than the everyday world of household goods, automobiles, or structural beams. In contrast, it is larger than the scale of atomic spacing by about four orders of magnitude, and it is larger even than the scale of dislocation spacing by approximately two orders of magnitude. Thus, even though the scale of an adiabatic shear band is well within the world of metallurgy and materials science, it does not seem unreasonable to apply standard continuum mechanics.

2.1 Balance Laws

Balance laws may be thought of as a form of accounting in which the total amount is found by keeping track of everything that flows in or out. Suppose that one is interested in some physical quantity that is contained within an arbitrary, but well defined, and fully enclosed volume. Then it is virtually a truism to say that the rate of change of the total amount of the quantity within the volume must arise from one of three causes. Either the physical quantity itself flows through the surface containing the volume, or there is a source of supply distributed over the surface of the volume, or there is a source of supply distributed throughout the interior of the volume. Written as an integral equation, this idea may be expressed as (e.g., see Truesdell and Toupin 1960 or Chadwick 1976)

$$\frac{d}{dt} \int_{\mathcal{V}} f \, dv = \oint_{\mathcal{S}} f (u_n - \mathbf{v} \cdot \mathbf{n}) \, ds + \oint_{\mathcal{S}} g \, ds + \int_{\mathcal{V}} h \, dv. \qquad (2.1)$$

In Equation (2.1) $\mathcal{V}$ is an arbitrarily moving volume in physical space and $\mathcal{S}$ is its surface. Most often the surface is imagined either to be fixed in the material and moving with the material particles or to be fixed in space so that the material flows through the stationary volume. The first case is called a material volume and the second a control volume, but any other arbitrary choice is possible. The physical quantity in question is represented by the volume density, $f(\mathbf{x}, t)$, which is imagined as being distributed throughout the volume and is associated with some material particle. Then $\int_{\mathcal{V}} f \, dv$ is the total amount of the quantity in $\mathcal{V}$, and the left-hand side of the equation represents the rate of change of that total. The first term on the right-hand side represents the rate at which the quantity is physically transported through the surface $\mathcal{S}$, where u_n is the outward speed of the surface itself in the direction of its exterior normal vector, $\mathbf{n}$, and $\mathbf{v}$ is the velocity of the physical particles that carry the quantity. Thus,

$(u_n - \boldsymbol{v} \cdot \boldsymbol{n})$ is the relative normal speed between the surface and the material. Finally g and h are surface and volume sources, respectively. So far the equation is a tautology. For it to become a statement of a physical law, specific meaning must be assigned to the quantities f, g, and h, which may be scalars, vectors, or tensors.

2.1.1 Balance of Mass

In Newtonian mechanics, mass is neither created nor destroyed. Then with the mass density ρ in place of f, and with $g = h = 0$, the balance of mass is written

$$\frac{d}{dt}\int_V \rho\,dv = \oint_S \rho(u_n - \boldsymbol{v} \cdot \boldsymbol{n})\,ds. \tag{2.2}$$

If the volume is a material volume, the right-hand side vanishes because $u_n - \boldsymbol{v} \cdot \boldsymbol{n} = 0$ everywhere on the surface by definition, but if the volume is a control volume, then $u_n = 0$ and the equation becomes $\frac{d}{dt}\int_V \rho\,dv + \oint_S (\boldsymbol{v} \cdot \boldsymbol{n})\rho\,ds = 0$. The second term is known as the Reynolds transport term.

Because the balance law holds for arbitrary volumes, it may also be expressed in differential form. It is possible to derive the desired differential equation by using arbitrary volume elements, but perhaps the simplest way is to consider only arbitrary control volumes that are fixed in space. Then the time derivative passes inside the volume integral, and after use of the divergence theorem on the surface integral, the balance law for mass becomes

$$\int_V [\partial_t \rho + \mathrm{div}(\rho\,\boldsymbol{v})]\,dv = 0. \tag{2.3}$$

Because the volume is arbitrary, the integrand must vanish, and the differential form of the law of conservation of mass becomes

$$\partial_t \rho + \mathrm{div}(\rho\,\boldsymbol{v}) = 0. \tag{2.4}$$

With the material time derivative defined as the time rate of change as seen by an observer who moves with the velocity of the material or $\dot{\rho} = \rho_t + \boldsymbol{v} \cdot \mathrm{grad}\,\rho$, Equation (2.4) may also be written as

$$\dot{\rho} + \rho\,\mathrm{div}\,\boldsymbol{v} = 0. \tag{2.5}$$

2.1.2 Balance of Momentum

When applied to a continuum, Newton's second law of motion takes the form

$$\frac{d}{dt}\int_V \rho\,\boldsymbol{v}\,dv = \oint_S \rho\,\boldsymbol{v}(u_n - \boldsymbol{v} \cdot \boldsymbol{n})\,ds + \oint_S \boldsymbol{t}\,ds + \int_V \rho\,\boldsymbol{b}\,dv, \tag{2.6}$$

where now f has been replaced by ρv, the density of linear momentum; g has been replaced by t, the density of surface tractions, and h has been replaced by ρb, the density of body forces. According to the usual tetrahedron argument, the surface tractions are related to the Cauchy stress tensor, T, by the linear mapping $t = Tn$ ($t_i = T_{ij}n_j$ in Cartesian tensor notation). After use has once again been made of the divergence theorem applied to an arbitrary control volume, the differential equation that corresponds to Equation (2.6) may be written as

$$\partial_t(\rho v) + \operatorname{div}(\rho v \otimes v) = \operatorname{div} T + \rho b. \tag{2.7}$$

Because the left-hand side of Equation (2.7) may be rewritten as $(\dot{\rho} + \rho \operatorname{div} v)v + \rho \dot{v}$, after use of Equation (2.5) the differential form of the law of conservation of linear momentum becomes

$$\operatorname{div} T + \rho b = \rho \dot{v}, \tag{2.8}$$

where $\dot{v}$ indicates the material time derivative of particle velocity.

2.1.3 Balance of Angular Momentum

Assuming that there are neither surface couples nor body couples in the material, one may write the integral form for the balance of angular momentum as

$$\frac{d}{dt} \int_{\mathcal{V}} x \times (\rho v) \, dv = \oint_{\mathcal{S}} [x \times (\rho v)] (u_n - v \cdot n) \, ds + \oint_{\mathcal{S}} x \times t \, ds$$

$$+ \int_{\mathcal{V}} x \times (\rho b) \, dv. \tag{2.9}$$

In Equation (2.9) the angular momentum and moments are all calculated relative to a fixed origin of coordinates, and the multiplication symbol $\times$ denotes the usual cross product of vectors. The same approach as used in the last two subsections again yields a differential equation, but because of the balance of mass and balance of momentum in Equations (2.4) and (2.8), only one term survives. In the notation of Cartesian tensors, the reduced form of Equation (2.9) is

$$\varepsilon_{ijk} T_{kj} = 0, \tag{2.10}$$

which is to say, the Cauchy stress tensor is symmetric.

2.1.4 Balance of Energy

A material particle may carry internal energy as a result of its state of deformation or its thermal state. It may also carry kinetic energy as a result of its state of motion. To express the balance of energy, Equation (2.1) now becomes

$$\frac{d}{dt}\int_{\mathcal{V}}\rho\left(e+\frac{1}{2}v^2\right)dv = \oint_{\mathcal{S}}\rho\left(e+\frac{1}{2}v^2\right)(u_n - \boldsymbol{v}\cdot\boldsymbol{n})\,ds + \oint_{\mathcal{S}}(\boldsymbol{v}\cdot\boldsymbol{t}-q)\,ds$$

$$+\int_{\mathcal{V}}\rho\,(\boldsymbol{v}\cdot\boldsymbol{b}+r)\,dv, \tag{2.11}$$

where e is the internal energy per unit mass, q is the heat flux (rate of heat transfer per unit area) leaving the volume, and r is an energy source per unit mass. The heat flux through a surface depends on the orientation of the surface; that is to say, it depends on the normal vector $\boldsymbol{n}$, but because there are two possible choices for $\boldsymbol{n}$ on a given element of surface, and because the physical direction of heat flux cannot depend on the choice, it must be that $q(\boldsymbol{n})=-q(-\boldsymbol{n})$. Then an argument similar to the tetrahedron argument for stress shows that q depends on the normal to the surface through the relationship $q=\boldsymbol{q}\cdot\boldsymbol{n}$, where $\boldsymbol{q}$ is the heat flux vector. Again appeal to a control volume with $u_n = 0$ and use of the divergence theorem leads to the differential equation

$$(\dot{\rho}+\rho\,\mathrm{div}\,\boldsymbol{v})\,(e+1/2v^2)+\boldsymbol{v}\cdot(\rho\dot{\boldsymbol{v}}-\mathrm{div}\,\boldsymbol{T}-\rho\boldsymbol{b})$$

$$+\rho\dot{e}=-\mathrm{div}\,\boldsymbol{q}+(\mathrm{grad}\,\boldsymbol{v}):\boldsymbol{T}+\rho r. \tag{2.12}$$

Because of the balance of mass (2.5) and the balance of momentum (2.8), only the last line survives. In summary, the three balance laws for mass, momentum, and energy are given by

$$\dot{\rho}+\rho\,\mathrm{div}\,\boldsymbol{v} = 0,$$

$$\rho\dot{\boldsymbol{v}}-\mathrm{div}\,\boldsymbol{T}-\rho\boldsymbol{b} = 0, \tag{2.13}$$

$$\rho\dot{e}+\mathrm{div}\,\boldsymbol{q}-(\mathrm{grad}\,\boldsymbol{v}):\boldsymbol{T}-\rho r = 0.$$

2.1.5 The Clausius–Duhem Inequality

In continuum mechanics the second law of thermodynamics is often expressed through the Clausius–Duhem inequality. If it is first assumed either that the internal entropy density per unit mass, η, is a primitive quantity requiring no further definition or explanation, or that it is a quantity that may be defined in operational terms, perhaps through cyclic processes, then the inequality for arbitrarily moving volumes may be written as

$$\frac{d}{dt}\int_{\mathcal{V}}\rho\eta\,dv \geq \oint_{\mathcal{S}}\rho\eta(u_n-\boldsymbol{v}\cdot\boldsymbol{n})\,ds + \oint_{\mathcal{S}}-\frac{\boldsymbol{q}\cdot\boldsymbol{n}}{T}\,ds + \int_{\mathcal{V}}\frac{\rho r}{T}\,dv, \tag{2.14}$$

where T is the absolute temperature. Proceeding as before by making use of the concept of a control volume and taking account of (2.5), one can derive the equivalent differential inequality:

$$\rho\dot{\eta} \geq -\operatorname{div}\left(\frac{q}{T}\right) + \frac{\rho r}{T}. \tag{2.15}$$

So far there are two scalar equations, one vector equation with three independent components, and a scalar inequality for a total of five equations and one inequality in all. However, even with the body force, b, and the heat source, r, prescribed arbitrarily, there are four scalar fields (ρ, e, η, T), two vector fields (v, q) with three independent components each, and a symmetric tensor field T with six independent components, for a total of sixteen independent components that must be determined to obtain a solution to any three-dimensional problem. This is a familiar situation in continuum mechanics, and one that must be solved by specifying constitutive laws. That is to say, the material response to mechanical and thermal disturbances must be completely described so that the system is no longer underdetermined.

The key to proceeding further is to give precise meaning to the phrase "mechanical and thermal disturbance," because all continuum theories require a careful description of the kinematic, thermal, and internal variables that are believed to be fundamental in specifying the state of the material. For example, in a thermoelastic material, the fundamental kinematic variable is the deformation gradient, $F = \partial x/\partial X$ ($F_{i\alpha} = \partial x_i/\partial X_\alpha = x_{i,\alpha}$ in Cartesian components), where x is the current location of a particle whose location in a reference configuration (usually stress free) was X, and the fundamental thermal variable is the specific entropy, η. In this case modern continuum mechanics provides a satisfactory structure that has gained wide acceptance.

2.2 Thermoelasticity

A brief review of thermoelasticity will provide useful perspective and insight before an attempt is made to pose a reasonably general version of plasticity. The material in the following section is standard and may be found in greater detail in Carlson (1972).

2.2.1 Objectivity and Implications of the Clausius–Duhem Inequality

In an ideal thermoelastic material it is assumed that the internal energy is a function only of the deformation gradient and the specific entropy, $e = e(F, \eta)$, and because the density of the material cannot become infinite, it must follow that $\det F \neq 0$ in any deformation. If the material is subjected to a rigid rotation, Q, after the original deformation has been applied, there is no further rearrangement of the internal particles relative to one another. Therefore, there can be no additional internal forces induced and no additional change in the internal energy. After the rotation, the total deformation gradient is $F^* = QF$, but the energy must retain the same value of $e(F^*, \eta) = e(F, \eta)$ for every rigid rotation. By the polar decomposition theorem (e.g., see Truesdell and Toupin 1960 or Ogden 1984), the deformation gradient F may always be uniquely decomposed into a pure stretch U, which is symmetric and positive definite, followed by a rigid rotation R, that is, $F = RU$. Because the deformation gradients F and F^* are energetically equivalent, that is $e(QRU, \eta) = e(RU, \eta)$ for every Q, in particular they are equivalent for $Q = R^T$ where the superscript denotes the transpose. Because $R^T R = RR^T = I$, the identity tensor, it must be that the energy depends only on the stretch, U. That is to say, $e(F, \eta) = e(U, \eta)$. This result is one consequence of the principle of material frame indifference. It states in a formal way the sensible idea that internal energy depends on the stretch in the material as measured relative to a suitable reference state (usually assumed to be unstressed) and not at all on the final orientation of the specimen in space. Stated another way, it is impossible to change the stored energy simply by making measurements in a different frame of reference.

The Clausius–Duhem inequality may be used to relate other field quantities to the internal energy. After the third equation of (2.13) has been substituted into (2.15) so as to eliminate $-\mathrm{div}\, q + \rho r$, followed by a rearrangement of terms, the entropy inequality becomes

$$\rho(\dot{e} - T\dot{\eta}) - T : \mathrm{grad}\, v \leq -\frac{q \cdot \mathrm{grad}\, T}{T}, \tag{2.16}$$

where v is the particle velocity relative to a spatial frame of reference, $v = \dot{x}$. With the chain rule applied to the internal energy, (2.16) becomes

$$\rho(e_\eta - T)\dot{\eta} + (\rho e_F - TF^{-T}) : \dot{F}^T \leq -\frac{q \cdot \mathrm{grad}\, T}{T}, \tag{2.17}$$

where F^{-T} is the inverse transpose of F and subscripts on e denote partial differentiation with respect to the argument indicated; in component form, $(TF^{-T}) : \dot{F}^T = T_{ij} X_{\alpha,j} \dot{x}_{i,\alpha}$.

In thermoelasticity the constitutive assumption is made that the stress and temperature, like the internal energy, also depend only on the deformation gradient and the specific entropy. If the heat flux does not depend on rates of entropy or deformation, and if inequality (2.17) is never to be violated for any accessible state, then the stress and the temperature must be obtainable from the internal energy:

$$T = \rho e_F F^T \quad (\text{or } T_{ij} = \rho e_{F_{i\alpha}} F_{j\alpha}),$$
$$T = e_\eta. \tag{2.18}$$

Furthermore, because the left-hand side of (2.17) now vanishes, the inequality reduces to

$$-\frac{\mathbf{q} \cdot \operatorname{grad} T}{T} \geq 0. \tag{2.19}$$

That is to say, because temperature is positive, the heat flux vector cannot make an acute angle with the temperature gradient. If the heat flux obeys Fourier's law, $\mathbf{q} = -\mathbf{K} \operatorname{grad} T$, then (2.19) requires that $(\operatorname{grad} T) \cdot (\mathbf{K} \operatorname{grad} T) \geq 0$. That is to say, $\mathbf{K}$ is at least positive semidefinite.

The energy equation in (2.13) may also be rewritten in expanded form as

$$\rho e_\eta \dot{\eta} + \rho e_F : \dot{\mathbf{F}}^T = -\operatorname{div} \mathbf{q} + \rho r + \mathbf{T} : (\operatorname{grad} \mathbf{v}), \tag{2.20}$$

but with the aid of the chain rule (which in this case is clearer in component form) and the first equation of (2.18), $\rho e_{F_{i\alpha}} \dot{x}_{i,\alpha} = \rho e_{F_{i\alpha}} \dot{x}_{i,j} x_{j,\alpha} = T_{ij} \dot{x}_{i,j}$. Then with the second equation of (2.18) the balance of energy reduces to

$$\rho T \dot{\eta} = -\operatorname{div} \mathbf{q} + \rho r. \tag{2.21}$$

To complete the reduction of the energy equation one must interpret the left-hand side of (2.21) in terms of readily understood physical quantities. To this end it is helpful first to introduce alternate measures of stress and strain and then to introduce the Helmholz free energy and the Gibbs function.

2.2.2 Helmholz and Gibbs Functions

It was noted above that the internal energy may depend on the deformation gradient only through the stretch tensor $\mathbf{U}$, but rather than using the stretch directly it is customary to use some measure of strain that vanishes in the reference state in which the stretch reduces to the identity. One such suitable and commonly used measure is $\mathbf{E} = 1/2(\mathbf{F}^T \mathbf{F} - \mathbf{I}) = 1/2(\mathbf{U}^2 - \mathbf{I})$, but many other measures are also possible (e.g., see Ogden 1984). Because the Cauchy stress is given by the first equation of (2.18), it may also be represented by

$$\mathbf{T} = \rho e_F \mathbf{F}^T = \rho \mathbf{F} e_E \mathbf{F}^T \quad \left(\text{or } T_{ij} = \rho \frac{\partial e}{\partial F_{i\alpha}} F_{j\alpha} = \rho \frac{\partial e}{\partial E_{\alpha\beta}} F_{i\alpha} F_{j\beta}\right). \tag{2.22}$$

Two other stress measures may be defined by

$$\mathbf{T}^R \equiv \frac{\rho_0}{\rho} \mathbf{T} \mathbf{F}^{-T} \quad \text{and} \quad \tilde{\mathbf{T}} \equiv \mathbf{F}^{-1} \mathbf{T}^R = \frac{\rho_0}{\rho} \mathbf{F}^{-1} \mathbf{T} \mathbf{F}^{-T}, \tag{2.23}$$

where ρ_0 is the density in the initial reference state. These stress tensors are related to the internal energy by

$$T^R = \rho_0 e_F \quad \left(\text{or } T_{i\alpha}^R = \rho_0 e_{F_{i\alpha}}\right),$$
$$\tilde{T} = \rho_0 e_E \quad \left(\text{or } \tilde{T}_{\alpha\beta} = \rho_0 e_{E_{\alpha\beta}}\right). \tag{2.24}$$

The tensor defined by the first equation of (2.23) and (2.24) is unsymmetric and is known as the first Piola–Kirchhoff stress tensor; the tensor defined by the second equation of (2.23) and (2.24) is symmetric and is known as the second Piola–Kirchhoff stress tensor. Both refer to force intensities as measured per unit reference area. With these stress measures the stress power per unit volume may be written alternatively as

$$T : \text{grad } v = \frac{\rho}{\rho_0} T^R : \dot{F}^T = \frac{\rho}{\rho_0} \tilde{T} : \dot{E} \quad \left(\text{or } T_{ij}\dot{x}_{i,j} = \frac{\rho}{\rho_0} T_{i\alpha}^R \dot{x}_{i,\alpha} = \frac{\rho}{\rho_0} \tilde{T}_{\alpha\beta} \dot{E}_{\alpha\beta}\right). \tag{2.25}$$

Because the expression $\rho \det F = \rho_0$ expresses the conservation of mass, as well as the first of (2.13), and the current volume is related to the reference volume by $dv = (\det F)\, dV$, the stress power in a volume dv may also be written as $\text{tr}(TD)\, dv = \text{tr}(T^R \dot{F}^T)\, dV = \text{tr}(\tilde{T}\dot{E})\, dV$, where $\text{tr}(\cdot)$ indicates the trace operation, dV is the volume element in the reference state, and the rate of stretching tensor, D, has Cartesian components $D_{ij} = 1/2(\dot{x}_{i,j} + \dot{x}_{j,i})$.

With the strain measure E and the stress measure $\tilde{T}$, Equation (2.18) may be rewritten as

$$e_E(E, \eta) = \rho_0{}^1 \tilde{T},$$
$$e_\eta(E, \eta) = T. \tag{2.26}$$

If it is assumed that these two expressions may be inverted, so that strain and entropy may be expressed as functions of stress and temperature

$$E = \bar{E}(\tilde{T}, T),$$
$$\eta = \bar{\eta}(\tilde{T}, T), \tag{2.27}$$

then the internal energy may also be expressed as a function of stress and temperature. That is,

$$e[\bar{E}(\tilde{T}, T), \bar{\eta}(\tilde{T}, T)] = \bar{e}(\tilde{T}, T). \tag{2.28}$$

Similarly, if only the second equation of (2.26) is inverted, the energy becomes

$$e[E, \hat{\eta}(E, T)] = \hat{e}(E, T). \tag{2.29}$$

It may be that there are multiple branches so that the inversion that results in (2.27) is not unique, but for present purposes it is enough to assume that the inversion is locally unique and may be extended uniquely until branch points occur.

The dependence of the internal energy on strain and entropy leads to the differential relation $\dot{e} = T\dot{\eta} + \rho_0^{-1}\tilde{T}:\dot{E}$, which after rewriting leads to the two relationships

$$\frac{d}{dt}(e - T\eta) = -\eta\dot{T} + \rho_0^{-1}\tilde{T}:\dot{E},$$

$$-\frac{d}{dt}\left(e - T\eta - \rho_0^{-1}\tilde{T}:E\right) = \eta\dot{T} + \rho_0^{-1}E:\dot{\tilde{T}}. \tag{2.30}$$

The terms in parentheses on the left-hand sides of (2.30) define the Helmholz free energy $\psi = e - T\eta$ and the Gibbs function $g = -e + T\eta + \rho_0^{-1}\tilde{T}:E$. If the inversions that led to (2.28) and (2.29) are unique, then the functions $\psi = \psi(E, T)$ and $g = g(\tilde{T}, T)$ are also unique and have derivatives

$$\psi_E = \rho_0^{-1}\tilde{T} = \rho_0^{-1}\hat{T}(E, T),$$

$$\psi_T = -\eta = -\hat{\eta}(E, T),$$

$$g_{\tilde{T}} = \rho_0^{-1}E = \rho_0^{-1}\bar{E}(\tilde{T}, T),$$

$$g_T = \eta = \bar{\eta}(\tilde{T}, T). \tag{2.31}$$

In (2.31) the caret and the overbar are used to emphasize the functional dependencies.

The mixed second derivatives of ψ and g lead to Maxwell relations. Because $\psi_{TE} = \psi_{ET}$ and $g_{T\tilde{T}} = g_{\tilde{T}T}$, from (2.31) it follows that

$$\hat{T}_T = -\rho_0\hat{\eta}_E,$$

$$\bar{E}_T = \rho_0\bar{\eta}_{\tilde{T}}. \tag{2.32}$$

2.2.3 Specific Heat, Thermal Stress, and Thermal Expansion

The specific heats of the material at constant strain and constant stress are defined by

$$c_E = \partial\hat{e}(E, T)/\partial T,$$

$$c_{\tilde{T}} = \{[\partial\bar{e}(\tilde{T}, T)]/\partial T\}, \tag{2.33}$$

which are generalizations of the specific heats in a fluid at constant volume and constant pressure, respectively. In (2.33) the subscripts on the left-hand sides only serve to distinguish the two specific heats and do not denote partial differentiation, of course. From the definitions of the Helmholz and Gibbs functions the specific heats may also be expressed as

$$c_E = \frac{\partial}{\partial T}(\psi + T\hat{\eta}) = T\frac{\partial\hat{\eta}}{\partial T} = -T\psi_{TT},$$

$$c_{\tilde{T}} = \frac{\partial}{\partial T}\left(-g + T\bar{\eta} + \rho_0^{-1}\tilde{T}:\bar{E}\right) = T\frac{\partial\bar{\eta}}{\partial T} + \rho_0^{-1}\tilde{T}:\frac{\partial\bar{E}}{\partial T} = Tg_{TT} + \tilde{T}:g_{\tilde{T}T}.$$

$$(2.34)$$

The definitions and notation are now in place to interpret the left-hand side of (2.21). The form of the result depends on whether the entropy is derived from the Helmholz or the Gibbs function. That is to say, the result depends on the choice of the independent constitutive variables. We have

$$\dot{\eta} = \hat{\eta}_T\dot{T} + \hat{\eta}_E:\dot{E} = \bar{\eta}_T\dot{T} + \bar{\eta}_{\tilde{T}}:\dot{\tilde{T}} \tag{2.35}$$

and consequently Equation (2.21), after use of (2.34), (2.35), and the appropriate Maxwell relation, reduces to either of two versions for the balance of energy:

$$\rho c_E\dot{T} = \mathrm{div}(\boldsymbol{K}\,\mathrm{grad}\,T) + \frac{\rho}{\rho_0}T\frac{\partial\tilde{T}}{\partial T}:\dot{E} + \rho r,$$

$$\rho\left(c_{\tilde{T}} - \rho_0^{-1}\tilde{T}:\frac{\partial E}{\partial T}\right)\dot{T} = \mathrm{div}(\boldsymbol{K}\,\mathrm{grad}\,T) - \frac{\rho}{\rho_0}T\frac{\partial E}{\partial T}:\dot{\tilde{T}} + \rho r.$$

$$(2.36)$$

Both versions of (2.36) show thermoelastic coupling. The term $\partial\tilde{T}/\partial T$ in the first equation of (2.36) is called the coefficient of thermal stress, because it represents the rate at which stress is induced as the temperature is changed with the strain held constant. The term $\partial E/\partial T$ in the second equation of (2.36) is called the coefficient of thermal expansion, because it represents the rate at which strain is induced as the temperature is changed with stress held constant. Both coefficients are tensors in general but reduce to scalars in isotropic materials.

Further relationships among the specific heats, coefficients of thermal stress, and coefficients of thermal expansion may be found by differentiation of various identities. For example, because $\hat{\eta}(\boldsymbol{E}, T) = \bar{\eta}(\tilde{T}, T)$ and $\boldsymbol{E} = \rho_0^{-1}\partial g(\tilde{T}, T)/\partial\tilde{T}$, when the entropy is differentiated with respect to temperature with stress held constant and when a Maxwell relation is used, it is straightforward to work out that

$$c_{\tilde{T}} - c_E = \frac{1}{\rho_0}\left(\tilde{T} - T\frac{\partial\tilde{T}}{\partial T}\right):\frac{\partial E}{\partial T} \tag{2.37}$$

Similarly, because $\tilde{T} = \rho_0^{-1}\partial\psi[\boldsymbol{E}(\tilde{T}, T), T]/\partial\boldsymbol{E}$, again differentiation with respect to temperature with stress held constant and use of a Maxwell relation leads to a relationship between the coefficients of thermal stress and thermal expansion:

$$0 = \frac{\partial \tilde{\boldsymbol{T}}}{\partial \boldsymbol{E}} : \frac{\partial \boldsymbol{E}}{\partial T} + \frac{\partial \tilde{\boldsymbol{T}}}{\partial T}. \tag{2.38}$$

The term $\partial \tilde{\boldsymbol{T}}/\partial \boldsymbol{E} = \rho_0 \partial^2 \psi /\partial \boldsymbol{E}^2$ represents the fourth-order tensor of isothermal elastic moduli. With the aid of (2.38) the relation between specific heats becomes

$$c_{\tilde{T}} - c_E = \frac{1}{\rho_0} \tilde{\boldsymbol{T}} : \frac{\partial \boldsymbol{E}}{\partial T} + T \frac{\partial \boldsymbol{E}}{\partial T} : \frac{\partial^2 \psi}{\partial \boldsymbol{E}^2} : \frac{\partial \boldsymbol{E}}{\partial T}. \tag{2.39}$$

Questions of material symmetry or changes of reference configuration will not be discussed in this section. Chapter 3, Section 3.2 contains some discussion about changes of reference configurations by rigid rotations, as applied to finite plasticity. The interested reader can find complete accounts for finite elasticity in Ogden (1984).

3

Thermoplasticity

The material in the preceding chapter is standard in theoretical thermoelasticity. It has been examined in some detail because it may be argued that it is also fundamental for plastically deforming solids. A typical crystalline or polycrystalline solid deforms elastically under small stresses or strains, but eventually, according to conventional views, the lattice becomes unstable and dislocations begin to form. This is the beginning of plastic deformation. As deformation continues, more dislocations form and existing dislocations move through the lattice. The net result is to add a certain amount of disorder or imperfection to the lattice. However, even at very large plastic strains most of the material remains crystalline. For example, suppose that in a heavily deformed material there are $\mathcal{O}(10^{12}\text{–}10^{13})$ dislocation lines cutting through 1 cm^2 (e.g., see Schaffer et al. 1995). This implies that the average spacing between dislocations is $\mathcal{O}(3\text{–}10)$ nm or approximately 10 to 40 atomic spacings or lattice parameters. Clearly most of the material must retain its crystalline nature, and if it is true that stress arises from the distortion of the crystal lattice in an elastic solid, then it must still be true in a plastically deformed solid. This simple observation is central to all further theoretical considerations.

Because the presence of a dislocation distorts the lattice in nearby regions, as shown schematically in Figure 3.1, panel a, elastic energy must be stored locally in its neighborhood. In fact, as is well known, a continuum representation of a dislocation results by considering a cut in the material, followed by a translation of one side of the cut surface relative to the other, and then followed by pasting the two sides of the surface back together again, as shown schematically in Figure 3.1, panel b. Self-equilibrated stresses will then be concentrated around the inner edge of the cut, and again it is clear that elastic energy will be stored locally near the dislocation. Thus either the lattice or the continuum picture results in broadly similar conclusions. Point defects, such as interstitial atoms

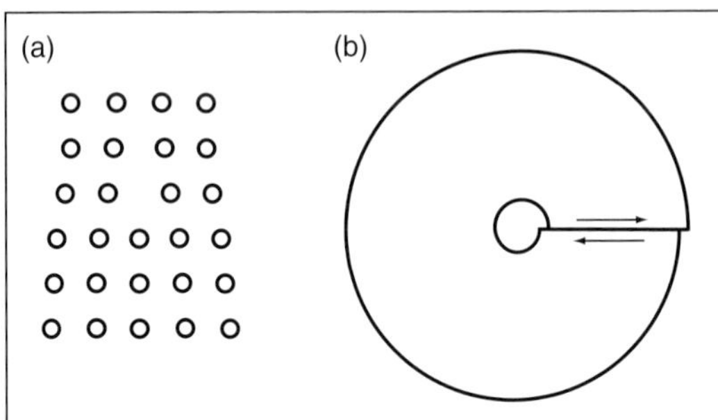

Figure 3.1. Schematic diagrams of a dislocation (panel a) in a lattice and (panel b) in a continuum. Either representation leads to a concentration of stored elastic energy near the core of the dislocation.

or vacancies, and surface defects, such as grain boundaries or stacking faults, similarly will cause a local distortion in the lattice and will have a similar effect on the internal energy.

To pursue this idea a bit further, let us imagine a small material volume in a thermoelastic material within which "uniform plastic deformation" is occurring, that is to say, within which a distribution of dislocations forms and changes as "plastic deformation" develops. According to our picture, there must be fluctuating stresses and strains within the volume element so that on a fine enough scale, there is nothing really uniform about the deformation at all. However, suppose we decompose both stress and strain into an average over the volume plus a fluctuation. Then an increment of elastic work per unit of current volume might be represented schematically as $\delta W = v^{-1} \int_v (\bar{\bar{T}} + \Delta\tilde{T}) : \delta(\bar{E} + \Delta E)\, dv$, where $\bar{\bar{T}}$ and $\bar{E}$ are average values over the current volume v (reference and current configurations are taken to be identical here), and δ and Δ signify the increment and the fluctuation, respectively. Because the average of the fluctuation within the volume is zero by definition, the increment of work may be written as

$$\delta W = \bar{\bar{T}} : \delta\bar{E} + v^{-1} \int_V \Delta\tilde{T} : \delta(\Delta E)\, dv \tag{3.1}$$

The first term obviously represents the elastic work done by the average stress and strain increment, and the second term represents the extra elastic work done by the fluctuations. In the language of materials science, the first term may be associated with "long-range forces," and the second with "short-range forces." To reiterate, the second term arises because of fluctuations in the elastic fields near dislocations or other disruptions to the lattice such as point defects, second phases, and the like. Alternatively it may also be associated with the stored energy of cold work.

The precise meaning of the various terms in (3.1) has deliberately been left vague, but it seems clear that if plastic deformation occurs, some theoretical account of the energetic consequences, not only of the average fields, but also of fluctuations near defects, should be given. In principle, it would seem to be possible to consider a plastically deformed body to be simply a special case of a thermoelastic solid with special rules adjoined to allow for the formation and propagation of individual dislocations. In practice, however, if approached deterministically such a program is wildly unrealistic because of the vast numbers of individual dislocations that must be accounted for. In contrast, the large numbers would seem to be an advantage for a statistical approach (e.g., see Kroner 1968; Sackett, Kelly, and Gillis 1972a, 1972b; or Ortiz and Popov 1982).

An alternate approach, and one that is in line with much other work in modern continuum mechanics, is to represent a plastically deformed material as being thermoelastic with respect to long-range forces, as suggested above, but in addition to represent the altered internal state by a finite set of internal variables, q_n, $n = 0, 1, 2, \ldots, N$, where according to Farshisheh and Onat (1974) and Adams, Boehler, Guidi, and Onat (1992), the q_n may be scalars or tensors of even order. It is intended that these internal variables be associated conceptually with the fluctuations and their effects on the average response. They are a reflection of the altered microstructure produced by plastic deformation and so may also be referred to as structural parameters. Two examples of such internal variables, or structural parameters, that are commonly used in classical plasticity are a scalar work hardening parameter and a tensorial back stress, but in this book only scalar internal variables will be discussed in any detail. In general terms now, the internal energy, the Helmholtz free energy, and the Gibbs energy all must depend on the internal variables, as well as their usual combinations of mechanical and thermal variables.

In the thermoelastic solid there is always a fixed reference state from which to measure deformations and to use in defining elastic strain as a fundamental kinematic quantity, but now the conveniently fixed reference state is lost because of plastic deformation. Stress and temperature are still available, however, so it seems most natural to take the point of view that in the plastically deformed state the Gibbs function is primary, and that the elastic strain and entropy are derived quantities. If the appropriate invertibility is assumed, and if the material is assumed to exhibit instantaneous thermoelasticity, all three thermodynamic potentials are still available (Scheidler and Wright 2001). This point of view is essentially identical with that taken in classical, small strain plasticity, but as we shall see, it is not quite as straightforward as it might seem at first glance when extended to finite deformations.

3.1 General Structure

Much of the following exposition is now fairly standard in the literature, which was recently reviewed critically by Cleja-Ţigoiu and Soós (1990). It should be noted, however, that the field has not yet reached a definitive form, and consequently there may be some significant departures here. The starting point is still the balance laws, as expressed in (2.13), and the entropy inequality, as expressed in (2.15). The presentation here is perhaps closest to that of Anand (1985) and Anand and Brown (1987).

3.1.1 Kinematics

First, however, it is necessary to refine the kinematics so as to include some measure of plastic deformation. The goal will be to find a procedure that decomposes the total deformation into elastic and plastic parts in as direct a manner as in classical, small strain plasticity. There, if the stress and the total strain are known, and if the elasticity relationship is known as well, then the elastic and plastic parts of the strain may easily be calculated. The well-known relationships in one dimension are

$$\varepsilon = \varepsilon^e + \varepsilon^p, \quad \varepsilon^e = \sigma/E, \quad \text{and therefore } \varepsilon^p = \varepsilon - \sigma/E. \qquad (3.2)$$

The elastic and plastic strains in this case are easily interpreted graphically as in Figure 3.2. Note that even without any knowledge of the uniaxial stress–strain

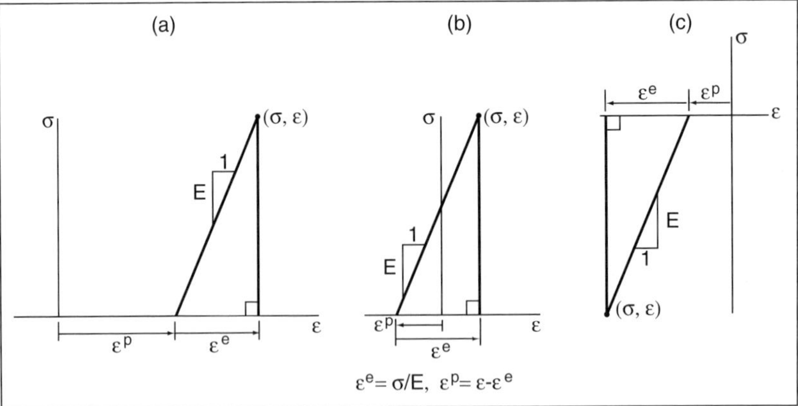

Figure 3.2. The decomposition of total strain into elastic and plastic parts is simple and direct in the classical, small strain theory, requiring only knowledge of the stress, the elastic constitutive relation (modulus E), and total strain at a point. The one-dimensional case is shown for three locations of the stress and total strain.

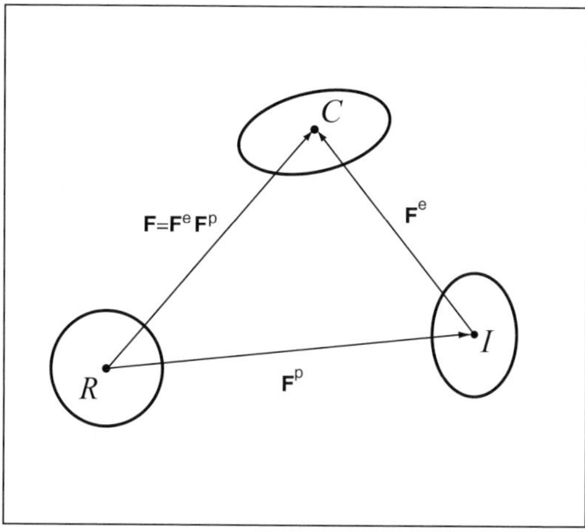

Figure 3.3. In finite deformation plasticity, it is customary to assume that total deformation, F, maps the reference configuration, R, onto the current configuration, C. The intermediate configuration, I, is defined conceptually by thermoelastic relaxation from C, and the plastic deformation maps R onto I. The net result is the product of elastic and plastic deformations, as proposed by Lee (1969).

curve itself or of the stress history, it is always possible to decompose the total strain unambiguously into its constituent parts. The nonlinear counterpart to (3.2) should be just as clear and unambiguous.

The situation in finite plasticity is, of course, much more complex, but a straightforward procedure is still possible. The usual decomposition for finite plasticity, also assumed here and shown in Figure 3.3, follows from the assumption that there is an unstressed, intermediate configuration, which can be defined conceptually by elastic and thermal relaxation from the current state of deformation. Thus, as noted by Lee (1969), the complete deformation gradient is given by the product of elastic and plastic deformation tensors, neither of which need be a gradient in itself.

$$F = F^e F^p. \tag{3.3}$$

Under rigid rotations, Q, the deformation transforms to $F^* = QF$, but it will be assumed that the plastic component of deformation, F^p, is invariant so that $F^{p^*} = F^p$.

Exactly how to define the intermediate configuration is a problem that will be ignored for the moment, but it can be proved that, even in the presence of continuing plastic deformation, the current state of stress and elastic deformation, as well as temperature and entropy, are related by thermodynamic potentials in a way that is entirely analogous to thermoelasticity (Scheidler and

Wright 2001; other authors either have assumed that the potential relations hold or have provided overly simplified proofs).

At this stage it is not necessary to assume that plastic deformation is isochoric, or volume preserving, so it will be supposed that the density in the intermediate configuration is ρ_R, and that it may change if there is further plastic deformation or recovery. Conservation of mass from the initial to the intermediate configuration and from the intermediate to the final configuration can be expressed by

$$\rho_R \det \boldsymbol{F}^p = \rho_0 \quad \text{and} \quad \rho \det \boldsymbol{F}^e = \rho_R. \tag{3.4}$$

Because $\det \boldsymbol{F} = \det \boldsymbol{F}^e \det \boldsymbol{F}^p$, these are completely consistent with overall conservation of mass.

It was observed by Clarebrough and coworkers in a series of papers (1952, 1955, 1956, 1957, 1962) that plastic volume change and the stored energy of cold work evolve simultaneously in several polycrystalline, fcc metals. Furthermore, Toupin and Rivlin (1960), using only second order, nonlinear elasticity, derived formulas that describe dimensional changes in a crystal that are due to the presence of a dislocation, and Wright (1982) showed that these formulas and data are fully consistent when supplemented by known, higher order, elastic moduli.

Therefore, because ρ_R can change with changing numbers of dislocations, or from the presence of voids, interstitial atoms, or substitutional atoms, and so on, it is logically required to be an internal variable itself, say $\rho_R = q_0$. At the very least, ρ_R must be regarded as a function of the other internal variables, $\rho_R = \rho_R(q_n)$.

The second Piola–Kirchhoff stress tensor may now be defined, relative to the Cauchy stress, $\boldsymbol{T}$, and the intermediate configuration, exactly as in thermoelasticity by

$$\tilde{\boldsymbol{T}} = \frac{\rho_R}{\rho} \boldsymbol{F}^{e^{-1}} \boldsymbol{T} \boldsymbol{F}^{e^{-T}}. \tag{3.5}$$

Note that ρ_R/ρ and $\boldsymbol{F}^e$ have replaced ρ_0/ρ and $\boldsymbol{F}$, respectively, from the original definition in (2.23). In the language of general continuum mechanics, the new definition corresponds exactly to a change of reference configuration from the original undeformed configuration to the intermediate configuration. Many authors have used this representation in finite plasticity, but usually with the additional assumption that the plastic deformation is isochoric (e.g., see Anand 1985).

3.1.2 Thermodynamic Potentials

As stated above, the Gibbs function is to be regarded as fundamental, so in each deformed state $g = g(\tilde{T}, T, q_n)$. When the q_n are all constant in a deformation and no additional plastic deformation occurs, the theory must reduce to the ideal thermoelastic case. This behavior is entirely analogous to the situation described in Chapter 2, but now the local reference density is ρ_R. Therefore, it is natural to identify the derivative of g with respect to stress as the elastic strain, and the derivative with respect to temperature as the entropy, even when the q_n are not constant in a deformation. Thus, it is assumed that

$$E^e \equiv \rho_R \frac{\partial g(\tilde{T}, T, q_n)}{\partial \tilde{T}},$$

$$\eta \equiv \frac{\partial g(\tilde{T}, T, q_n)}{\partial T},$$

(3.6)

where E^e is related to F^e by

$$E^e = 1/2\left(F^{e^T} F^e - I\right).$$

(3.7)

Equations (3.6) can also be derived under the simple and physically reasonable assumption that the material exhibits instantaneous thermoelasticity (Scheidler and Wright 2001).

The other two thermodynamic potentials may be defined by the same relations as before, again with ρ_R replacing ρ_0, and F^e replacing F:

$$\psi = g + \rho_R^{-1}\tilde{T}:E^e,$$

$$e = -g + T\eta + \rho_R^{-1}\tilde{T}:E^e.$$

(3.8)

Assuming the appropriate invertibility of (3.6), the functional dependencies of ψ and e follow from the assumed dependence of g on $(\tilde{T}, T, q_n)$, as will now be demonstrated.

The differential of g, when written out in full, is given by

$$dg = \rho_R^{-1}E^e : d\tilde{T} + \eta dT + \sum_{n=0}^{N} \frac{\partial g}{\partial q_n} dq_n,$$

(3.9)

so with (3.9) taken into account and with $\rho_R = q_0$, the differentials of (3.8) are written as

$$d\psi = \rho_R^{-1}\tilde{T}:dE^e - \eta dT - \left(\frac{\partial g}{\partial q_0} + \frac{1}{\rho_R^2}\tilde{T}:E^e\right)dq_0 - \sum_{n=1}^{N} \frac{\partial g}{\partial q_n} dq_n,$$

$$de = \rho_R^{-1}\tilde{T}:dE^e + Td\eta - \left(\frac{\partial g}{\partial q_0} + \frac{1}{\rho_R^2}\tilde{T}:E^e\right)dq_0 - \sum_{n=1}^{N} \frac{\partial g}{\partial q_n} dq_n.$$

(3.10)

Derivatives of the Gibbs function with respect to the q_n will be used to define a set of new entities, Q_n, as follows:

$$-\left(\frac{\partial g}{\partial q_0} + \frac{1}{\rho_R^2}\tilde{T}:E^e\right) \equiv \rho_R^{-1}Q_0,$$

$$-\frac{\partial g}{\partial q_n} \equiv \rho_R^{-1}Q_n, \quad 1 \le n \le N. \tag{3.11}$$

With these definitions the differentials of the three potentials become

$$de = \rho_R^{-1}\tilde{T}:dE^e + Td\eta + \rho_R^{-1}\sum_{n=0}^{N} Q_n dq_n,$$

$$d\psi = \rho_R^{-1}\tilde{T}:dE^e - \eta dT + \rho_R^{-1}\sum_{n=0}^{N} Q_n dq_n,$$

$$dg = \rho_R^{-1}E^e:d\tilde{T} + \eta dT - \rho_R^{-1}\left(Q_0 + \rho_R^{-1}\tilde{T}:E^e\right)d\rho_R - \rho_R^{-1}\sum_{n=1}^{N} Q_n dq_n. \tag{3.12}$$

Equation (3.12) is entirely consistent with the expected functional dependencies of e and ψ, namely

$$e = e(E^e, \eta, q_n),$$
$$\psi = \psi(E^e, T, q_n). \tag{3.13}$$

Furthermore, the partial derivatives of the potentials produce the expected results.

$$\frac{\partial e}{\partial E^e} = \rho_R^{-1}\tilde{T}, \quad \frac{\partial e}{\partial \eta} = T, \quad \frac{\partial e}{\partial q_n} = \rho_R^{-1}Q_n, \qquad n = 0, 1, \ldots, N,$$

$$\frac{\partial \psi}{\partial E^e} = \rho_R^{-1}\tilde{T}, \quad \frac{\partial \psi}{\partial T} = -\eta, \quad \frac{\partial \psi}{\partial q_n} = \rho_R^{-1}Q_n, \qquad n = 0, 1, \ldots, N,$$

$$\frac{\partial g}{\partial \tilde{T}} = \rho_R^{-1}E^e, \quad \frac{\partial g}{\partial T} = \eta, \quad \frac{\partial g}{\partial q_n} = -\rho_R^{-1}Q_n - \delta_n^0\rho_R^{-2}\tilde{T}:E^e, \quad n = 0, 1, \ldots, N,$$

$$\tag{3.14}$$

where $\delta_n^0 = 1$ if $n = 0$ and vanishes otherwise.

The scheme outlined here is identical to the ideal thermoelastic case when the internal structure is constant, and so it represents only a simple extension that it should still hold true as the structural parameters vary. The mixed second derivatives of g lead to the Maxwell relations

$$\frac{\partial Q_n}{\partial \tilde{T}} + \delta_n^0 \tilde{T} : \frac{\partial^2 g}{\partial \tilde{T}^2} = -\frac{\partial E^e}{\partial q_n},$$

$$\frac{\partial Q_n}{\partial T} + \delta_n^0 \frac{1}{\rho_R} \tilde{T} : \frac{\partial E^e}{\partial T} = -\rho_R \frac{\partial \eta}{\partial q_n}, \tag{3.15}$$

where δ_n^0 again is the Kronecker delta.

3.1.3 Entropy and Energy

The framework is now in place to reexamine the structure of the entropy inequality and the balance of energy, given the kinematics and potential functions just described. When the second equation of (3.8) is substituted into the entropy inequality (2.16), and when the derivatives in (3.14) are used, the entropy inequality reduces to

$$T : D - \frac{\rho}{\rho_R} \tilde{T} : \dot{E}^e - \frac{\rho}{\rho_R} \sum_{n=0}^{N} Q_n \dot{q}_n - \frac{q \cdot \nabla T}{T} \geq 0. \tag{3.16}$$

The first term in Equation (3.16) is the total stress power per unit current volume. According to the concept of field averages and fluctuations, the second term appears to be the rate of change of elastic energy that is due to mechanical power of the average stresses or long-range forces, whereas the third term represents the rate of elastic energy storage that is due to the fluctuations around dislocations or other short-range forces. Ordinarily the second term would be called the rate of storage of elastic energy or the elastic stress power, and because the third term, although also arising from elastic changes, only exists when the internal state is changing, it may be associated with the stored energy of cold work (and perhaps also with the localized elastic energy stored in the lattice around point defects).

In elastic deformation the first two terms of (3.16) cancel, and when there also are no structural changes, the third term vanishes, leaving the well-known inequality for heat conduction. It seems reasonable to assume that the heat inequality still holds in plastic deformation, as well, although strictly speaking it need not. Furthermore, the first three terms in (3.16) must satisfy the inequality themselves when the temperature gradient vanishes, and because they do not depend on the temperature gradient, they must always satisfy the inequality. The net result is that (3.16) is now split into two inequalities, the first involving only mechanical changes and the second only thermal quantities:

$$T:D - \frac{\rho}{\rho_R}\tilde{T}:\dot{E}^e - \frac{\rho}{\rho_R}\sum_{n=0}^{N} Q_n\dot{q}_n \geq 0,$$

$$-\frac{\boldsymbol{q}\cdot\nabla T}{T} \geq 0. \tag{3.17}$$

Closer examination of the first part of (3.17) shows that the first term may be rewritten in a more advantageous way. Recall that $D = 1/2(\dot{F}F^{-1} + F^{-T}\dot{F}^T)$, and thus by (3.3), (3.7), and (3.5), we have $\dot{F} = \dot{F}^e F^p + F^e \dot{F}^p$, $\dot{E}^e = 1/2(\dot{F}^{e^T} F^e + F^{e^T} \dot{F}^e)$, and $T = (\rho/\rho_R)F^e\tilde{T}F^{e^T}$, respectively. Then after it is recalled that $\operatorname{tr}AB = \operatorname{tr}BA = A:B = B:A$, it is only a short algebraic exercise to work out that

$$T:D = \frac{\rho}{\rho_R}\tilde{T}:\dot{E}^e + T:D^p, \tag{3.18}$$

where D^p, which may be called the plastic strain rate, is given by

$$D^p = 1/2\left(F^e \dot{F}^p F^{p^{-1}} F^{e^{-1}} + F^{e^{-T}} F^{p^{-T}} \dot{F}^{p^T} F^{e^T}\right). \tag{3.19}$$

Note that just as $\operatorname{tr}D = \operatorname{tr}\dot{F}F^{-1} = d/dt\,(\det F)/\det F = -\dot{\rho}/\rho$, in a similar way we have $\operatorname{tr}D^p = \operatorname{tr}\dot{F}^p F^{p^{-1}}$ from (3.19), and because $(\det F^p)\operatorname{tr}\dot{F}^p F^{p^{-1}} \equiv d/dt(\det F^p)$ and $\rho_R \det F^p = \rho_0$, conservation of mass in the intermediate configuration may be expressed as

$$\operatorname{tr}D^p = -\dot{\rho}_R/\rho_R. \tag{3.20}$$

Equations equivalent to (3.18) and (3.19) are commonly written in the literature of finite plasticity, but often with variations of the definitions. For example, Anand (1985) focuses attention on the term $\dot{F}^p F^{p^{-1}}$, and Bammann and Johnson (1987) consider $\dot{U}^p U^{p^{-1}}$ where $(U^p)^2 = F^{p^T} F^p$. The relationships among several alternative choices for the principal representation of plastic flow are discussed by Scheidler and Wright (2001).

Equation (3.18) now allows the first part of inequality (3.17) to be written as

$$T:D^p - \frac{\rho}{\rho_R}\sum_{n=0}^{N} Q_n\dot{q}_n \geq 0. \tag{3.21}$$

Inequality (3.21) states that the plastic stress power less the rate of change of internal energy that is due to changing internal variables cannot be negative.

Inequality (3.21) holds even if there is plastic volume change, but when it is recalled that $\dot{q}_0$ and $\operatorname{tr}D^p$ are both associated with plastic volume change, it becomes apparent that a better representation would follow from a decomposition of T and D^p into their spherical and deviatoric parts. Thus, with the stress deviator, S, and the deviator of the plastic rate, d^p, defined by the identities $T \equiv 1/3(\mathrm{tr}\,T)I + S$ and $D^p \equiv 1/3(\mathrm{tr}\,D^p)I + d^p$, inequality (3.21) with the aid of (3.20) may be rewritten as

$$S:d^p + (p - \rho Q_0)\frac{\dot{\rho}_R}{\rho_R} - \frac{\rho}{\rho_R}\sum_{n=1}^{N} Q_n \dot{q}_n \geq 0, \tag{3.22}$$

because pressure is given by $p = -1/3\,\mathrm{tr}\,T$. The combination of terms $p - \rho Q_0$ represents an effective pressure that does work against plastic volume change.

With the Clausius–Duhem inequality in the form of (3.22), it can clearly be seen that constitutive laws for the deviator of the plastic rate, the rate of plastic volume change, and the rate of change of all the other internal variables must be specified in order to complete the description of material behavior. The specification for d^p is called the flow law. In general the rate of plastic volume change must be specified by a constitutive law, but if it is assumed to be zero $(\mathrm{tr}\,D^p = 0)$, as in classical plasticity theories, then Q_0 may be an arbitrarily prescribed constraint force. Finally, the remaining rates of change of internal variables, $\dot{q}_n$, $n = 1, 2, \ldots N$, must also be specified by constitutive laws. The Clausius–Duhem inequality demands that all of these rates must be given in such a way that inequality (3.22) is never violated.

In the special case that the internal energy and the free energy do not depend directly on ρ_R, but ρ_R does depend on the other internal variables, then $Q_0 = 0$ and inequality (3.22) may be written as

$$S:d^p - \frac{\rho}{\rho_R}\sum_{n=1}^{N}\left(Q_n - \frac{p}{\rho}\frac{\partial\rho_R}{\partial q_n}\right)\dot{q}_n \geq 0. \tag{3.23}$$

The kinematics and potential functions that lead to (3.22) also affect the way in which conservation of energy from the third equation of (2.13) may be expressed.

$$\rho T\dot{\eta} = -\mathrm{div}\,q + \rho r + T:D^p - \frac{\rho}{\rho_R}\sum_{n=0}^{N} Q_n \dot{q}_n. \tag{3.24}$$

Because the specific entropy, η, has been obtained from a thermodynamic potential, the left-hand side of (3.24) may be expanded in exactly the same manner as in the purely thermoelastic case that resulted in (2.36), but now the internal variables must also be taken into account. The result is

$$\rho c_E \dot{T} = -\mathrm{div}\, \boldsymbol{q} + \rho r + \frac{\rho}{\rho_R} T \frac{\partial \tilde{\boldsymbol{T}}}{\partial T} : \dot{\boldsymbol{E}}^e + \boldsymbol{T} : \boldsymbol{D}^p - \frac{\rho}{\rho_R} \sum_{n=0}^{N} \left(Q_n - T \frac{\partial Q_n}{\partial T} \right) \dot{q}_n.$$

$$(3.25)$$

The last two terms in (3.24) or (3.25) may also be expanded as in (3.22), of course. The last term in the final sum may be called the thermoplastic term because it is analogous to the thermoelastic term. Both arise from the expansion of the material time derivative of entropy in (3.24). When the last two terms vanish, the balance of energy reduces exactly to the thermoelastic case, as in (2.36), but now the intermediate configuration for each material point serves as the local elastic reference configuration.

3.2 Yield, Plastic Flow, and Constitutive Equations

The central problem for plasticity is to give an account of the evolution of the plastic part of the deformation, and the first part of that task is to determine what exactly is meant by plastic deformation. Once the decomposition of the deformation gradient in (3.3) is known, the overall structure of the theory, as laid out in the last section, is well defined. In contrast, if only the total deformation gradient and the Cauchy stress are known, then the decomposition is unknown a priori and remains to be found. In the general case, the material will be elastically anisotropic, and the missing ingredient, not yet specified, is the orientation of the material in the intermediate configuration. That is to say, the elastic symmetries of the intermediate configuration must be precisely stated before it is possible to compute the stress–strain relationship.

Of course it is always possible to make a change of reference configuration for any elastic material (e.g., see Ogden 1984), but doing so changes the response function and the symmetry group in a corresponding way. Figure 3.4, panel a shows the general situation for finite elasticity when the change of reference configuration is produced by a simple rotation. This idea carries over to finite elastoplasticity, where it may be applied to the intermediate configuration.

3.2.1 The Intermediate Configuration and the Local Elastic
Reference Configuration

Simple rotations of the intermediate configuration produce changes of the local elastic reference configuration. For example, suppose that instead of the given elastic deformation, $\boldsymbol{F}^e$, it is desired to use $\boldsymbol{F}^{e^*} = \boldsymbol{F}^e \boldsymbol{P}^T$, where $\boldsymbol{P}$ is some arbitrary rotation, as shown in Figure 3.4, panel b. From the polar

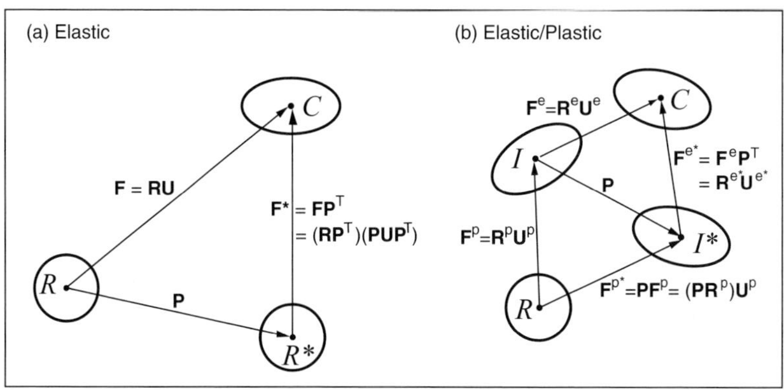

Figure 3.4. As shown in panel a, a finite elastic reference configuration, R, may be mapped by a rigid rotation, $\boldsymbol{P}$, onto another, equally valid, reference configuration, R^*. Similarly, as shown in panel b, the intermediate configuration, I, may also be mapped by a rigid rotation, $\boldsymbol{P}$, onto an alternate, but equally valid, intermediate configuration, I^*.

decomposition theorem, we have $\boldsymbol{F}^e = \boldsymbol{R}^e \boldsymbol{U}^e$, and therefore also $\boldsymbol{F}^{e*} = \boldsymbol{R}^e \boldsymbol{U}^e \boldsymbol{P}^T = (\boldsymbol{R}^e \boldsymbol{P}^T)(\boldsymbol{P}\boldsymbol{U}^e \boldsymbol{P}^T) = \boldsymbol{R}^{e*} \boldsymbol{U}^{e*}$. Because the final configuration of the material is the same in both cases, the elastic response to the deformation from the new reference configuration must deliver the same Cauchy stress as the original elastic deformation and response. It is easily worked out from (3.5) that the second Piola–Kirchhoff stress transforms in the same way as the elastic stretch under a change of reference configuration, so that $\tilde{\boldsymbol{T}}^* = \boldsymbol{P}\tilde{\boldsymbol{T}}\boldsymbol{P}^T$. Consequently, if $\tilde{\boldsymbol{T}} = \boldsymbol{\mathcal{T}}(\boldsymbol{U}^e)$ and $\tilde{\boldsymbol{T}}^* = \boldsymbol{\mathcal{T}}^*(\boldsymbol{U}^{e*})$, then the two response functions are related by

$$\boldsymbol{\mathcal{T}}^*(\boldsymbol{U}^{e*}) = \boldsymbol{P}\boldsymbol{\mathcal{T}}(\boldsymbol{P}^T \boldsymbol{U}^{e*} \boldsymbol{P})\boldsymbol{P}^T \tag{3.26}$$

for every elastic stretch $\boldsymbol{U}^{e*}$. In general, the response functions $\boldsymbol{\mathcal{T}}(\cdot)$ and $\boldsymbol{\mathcal{T}}^*(\cdot)$ are not the same unless $\boldsymbol{P}$ is a member of the symmetry group of the material. The relationship (3.26) simply says that to calculate the stress from the starred configuration, first rotate the stretch to the original configuration, then calculate with the known response function, and finally rotate back to the starred configuration. In other words, if there is a preferred, reference configuration for the elastic calculation (the unstarred configuration indicated by $\mathcal{I}$ in the figure, say) where the response function, including the symmetry group, is known, it may always be used to calculate stresses provided that its orientation is known relative to another unstressed, elastic reference configuration (the starred configuration indicated by $\mathcal{I}^*$ in the figure, say), which has been used for decomposition of the deformation. The key then is to know the response function, including the symmetry group, for one preferred elastic reference configuration and the rotation required to get there from the present elastic reference configuration.

These facts concerning change of elastic reference configuration (i.e., change of intermediate configuration) and material symmetry seem to have caused no end of confusion in the plasticity literature. For example, virtually all writers agree on using the decomposition $F = F^e F^p = R^e U^e R^p U^p$. From the preceding discussion, it seems obvious that the terms could be regrouped as $F = R^{e^*} U^{e^*} U^p = (R^e R^p)(R^{p^T} U^e R^p)U^p$, as favored by some authors, or as $F = U^{e^*} R^{p^*} U^p = (R^e U^e R^{e^T})(R^e R^p)U^p$, as favored by other authors, or in an infinity of other arbitrary ways, as well. Figure 3.4, panel b shows one such alternate choice for the intermediate configuration. Each rewriting and regrouping of terms only amounts to a change of reference configuration for the elasticity calculation, as may be seen from the figure. However, as stated above, whatever the choice, it is still necessary to know the response function, including the symmetry group, from one preferred configuration (not necessarily the one chosen for the decomposition of F) and also to know the rotation that is necessary to move from the configuration chosen for decomposition to the preferred configuration chosen for computation of stress. In that sense, a "plastic rotation" is generally required, and a constitutive equation must be given for its motion, say $\dot{R}^p R^{p^T} = \Omega(\tilde{T}, T, q_n)$, where Ω is an antisymmetric, tensor-valued function. Unfortunately, without some further physical motivation there does not seem to be a satisfactory way to prescribe this motion for a general anisotropic material.

A better approach to the general case, which also attempts to be faithful to the underlying crystal structure and the available slip mechanisms, is given by the application of polycrystalline plasticity (Hill and Rice 1972, Asaro 1982, Havner 1992, and others).

3.2.2 Elastically Isotropic Materials With Scalar Internal Variables Only

If the material is assumed to be elastically isotropic, with the corresponding principal stresses and elastic stretches being parallel to each other, the desired decomposition may be constructed in a simple and elegant way. In this case, $\tilde{T}$ is an isotropic, tensor-valued function of U^e and vice versa, by the assumed elastic isotropy and invertibility. Consequently, the Cauchy stress is an isotropic, tensor-valued function of the left elastic stretch, V^e, which transforms under rigid body motions in exactly the same way as the Cauchy stress itself. Thus, if Q is a rigid motion that transforms the deformation gradient as $F^* = QF$, the Cauchy stress transforms as $T^* = QTQ^T$, and similarly the left stretch transforms as $V^{e^*} = QV^e Q^T$. Furthermore, because the material has been assumed to have isotropic elastic symmetry, the elastic constitutive relation may be written $T = \mathcal{F}(V^e)$, which has the property $\mathcal{F}(QV^e Q^T) = Q\mathcal{F}(V^e)Q^T$. (Dependence on temperature and internal variables has been suppressed.) Generally in this case T and V^e are also coaxial. That is, a proper vector of one is also a proper vector of the other. Because the relationship between stress and stretch has been assumed to be invertible, then $V^e = \mathcal{F}^{-1}(T)$ is known, and the product, $V^{e-1}F$, may be uniquely decomposed by the polar decomposition theorem into the product of a rotation, R, and a pure stretch, denoted U^p. That is to say, for a given Cauchy stress the deformation gradient may be uniquely decomposed into the product

$$F = V^e R U^p, \tag{3.27}$$

where V^e is related to the stress, T, through the elastic constitutive relation.

Because R in (3.27) is fully determined by the preceding procedure, there is no reason to partition it further into elastic and plastic parts. Furthermore, a direct calculation using (3.27) shows that the total stress power does not depend on the rate of change of R. Demonstration of this last observation depends on the fact that T and V^e are coaxial and that $\dot{R}R^T$ is antisymmetric. A short calculation leads to the expression

$$\begin{aligned}
2T:D &= T:(\dot{F}F^{-1} + F^{-T}\dot{F}^T) \\
&= T:(\dot{V}^e V^{e-1} + V^{e-1}\dot{V}^e) + \dot{R}R^T:(V^{e-1}TV^e - V^e TV^{e-1}) \\
&\quad + T:(V^e R\dot{U}^p U^{p-1} R^T V^{e-1} + V^{e-1}RU^{p-1}\dot{U}^p R^T V^e).
\end{aligned} \tag{3.28}$$

The first term in (3.28) is the elastic stress power, the second vanishes because of coaxiality, and the third is the plastic stress power.

Consequently, the right elastic stretch tensor may be defined from the usual relation, $U^e = R^T V^e R$, and Equation (3.27) may also be written uniquely as

$$F = RU^e U^p. \tag{3.29}$$

The right stretch, U^e, so defined, together with the rotation, R, are guaranteed to deliver the Cauchy stress. In effect the rotation found uniquely in (3.27) has been assigned completely to F^e so that $F^p = U^p$. The net result for elastically isotropic materials with only scalar internal variables is that the plastic part of the total deformation may always be regarded as a pure stretch, which may be found uniquely, provided the Cauchy stress and the elastic constitutive relation are known. Conceptually this decomposition has the desired property that it is nearly as simple as (3.2).

If the total deformation and second Piola–Kirchhoff stress, rather than the Cauchy stress, are known, it is still possible to construct the decomposition (3.29) as follows (Scheidler, personal communication). Because it is now known that the decomposition exists and is unique for an isotropic material, and because with $\tilde{T}$ given, the elastic stretch, U^e, is known from the elastic constitutive relation, then from (3.29) U^p must satisfy $F^T F = U^2 = U^p U^{e^2} U^p$. After pre-multiplying and postmultiplying by the elastic stretch, the plastic stretch satis-fies $(U^e U^p U^e)^2 = U^e U^2 U^e$, and as a consequence

$$U^p = U^{e^{-1}}(U^e U^2 U^e)^{1/2} U^{e^{-1}}. \tag{3.30}$$

With (3.30) as a definition for U^p, it is easy to show that $FU^{p^{-1}} U^{e^{-1}}$ has an inverse equal to its transpose and therefore the product is a pure rotation. Again Equation (3.29) is the result.

3.2.3 Plastic Stretching in Elastically Isotropic Materials

Returning now to Equation (3.18), with the decomposition (3.29) one may write the total stress power as

$$T:D = T:(D^e + D^p), \tag{3.31}$$

where the elastic and plastic parts of the stretching tensor are given by

$$\begin{aligned}
D^e &= 1/2\, R(\dot{U}^e U^{e-1} + U^{e-1}\dot{U}^e)R^T, \\
D^p &= 1/2\, R(U^e \dot{U}^p U^{p-1} U^{e-1} + U^{e-1} U^{p-1}\dot{U}^p U^e)R^T.
\end{aligned} \tag{3.32}$$

From the forms of the two rates in (3.32) and from the structure of the decomposition of deformation it is clear that both D^e and D^p transform under rigid body motions in the same way as the Cauchy stress and therefore are objective rates.

Alternatively the total stress power may be written as

$$T:D = \frac{\rho}{\rho_R}\tilde{T}:(\tilde{D}^e + \tilde{D}^p), \tag{3.33}$$

where now the elastic and plastic parts of the stretching tensor have been transformed into the intermediate configuration where they are unaffected by changes of observer and are given by

$$\begin{aligned}
\tilde{D}^e &= 1/2\,(U^e\dot{U}^e + \dot{U}^e U^e) = \dot{E}^e, \\
\tilde{D}^p &= 1/2\bigl(U^{e^2}\dot{U}^p U^{p-1} + U^{p-1}\dot{U}^p U^{e^2}\bigr).
\end{aligned} \tag{3.34}$$

A comparison of (3.32) with (3.34) shows that the spatial and material forms of the elastic and plastic stretching tensors are related by

$$\begin{aligned}
D^e &= F^{e^{-T}}\tilde{D}^e F^{e^{-1}} = RU^{e^{-1}}\tilde{D}^e U^{e^{-1}} R^T, \\
D^p &= F^{e^{-T}}\tilde{D}^p F^{e^{-1}} = RU^{e^{-1}}\tilde{D}^p U^{e^{-1}} R^T.
\end{aligned} \tag{3.35}$$

There has been considerable discussion in the literature of finite plasticity concerning the proper choice of term to represent the "plastic rate of

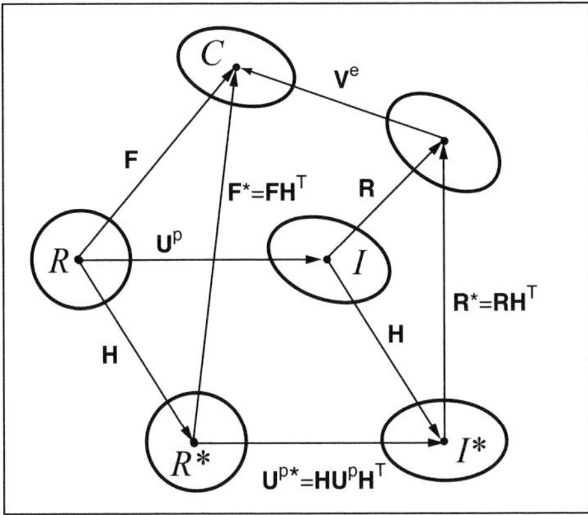

Figure 3.5. In finite plasticity the initial reference configuration may be mapped by a rigid rotation, **H**, onto an alternate reference configuration, R^*. Consequences that this initial rotation will have on the elastic-plastic decomposition are shown. The case for isotropic elasticity is shown.

deformation." Some authors have favored $\boldsymbol{D}^p$, some $\tilde{\boldsymbol{D}}^p$, and some still other forms (as discussed, e.g., in Lubliner 1990). Equation (3.35) suggests that it makes little theoretical difference which is favored because the various forms may be transformed back and forth at will. An examination of the concepts of isotropic material symmetry, plastic yield surface, and flow rule tends to reinforce the latter point of view.

Changes of the initial reference configuration may also be made. Thus, if $\boldsymbol{H}$ is a constant, proper orthogonal transformation that rotates the initial configuration before the deformation is applied, then the new deformation gradient transforms to $\boldsymbol{F}^* = \boldsymbol{F}\boldsymbol{H}^T$ (refer to Figure 3.5). Because the final configuration is unchanged, the Cauchy stress and the left elastic stretch tensor remain unchanged, and the new deformation gradient may be uniquely decomposed as $\boldsymbol{F}^* = \boldsymbol{V}^e\boldsymbol{R}^*\boldsymbol{U}^{p^*}$. However, from the unique decomposition of $\boldsymbol{F}$, $\boldsymbol{F}^*$ may be written as $\boldsymbol{F}^* = (\boldsymbol{V}^e\boldsymbol{R}\boldsymbol{U}^p)\boldsymbol{H}^T = \boldsymbol{V}^e(\boldsymbol{R}\boldsymbol{H}^T)(\boldsymbol{H}\boldsymbol{U}^p\boldsymbol{H}^T)$, which identifies the new components of $\boldsymbol{F}^*$ as

$$\boldsymbol{R}^* = \boldsymbol{R}\boldsymbol{H}^T \quad \text{and} \, \boldsymbol{U}^{p^*} = \boldsymbol{H}\boldsymbol{U}^p\boldsymbol{H}^T. \tag{3.36}$$

Furthermore, because $\boldsymbol{V}^e = \boldsymbol{R}\boldsymbol{U}^e\boldsymbol{R}^T$ by construction, and because $\boldsymbol{V}^e$ is unchanged by the initial rotation, it may also be written as $\boldsymbol{V}^e = (\boldsymbol{R}\boldsymbol{H}^T)(\boldsymbol{H}\boldsymbol{U}^e\boldsymbol{H}^T)(\boldsymbol{H}\boldsymbol{R}^T)$. Therefore, it may be concluded that

$$\boldsymbol{U}^{e^*} = \boldsymbol{H}\boldsymbol{U}^e\boldsymbol{H}^T \quad \text{and} \, \boldsymbol{V}^e = \boldsymbol{R}^*\boldsymbol{U}^{e^*}\boldsymbol{R}^{*T} \tag{3.37}$$

as expected.

Because the rotation, H, is constant, it is easily demonstrated with equations (3.36) and (3.37) substituted into (3.34) and (3.32) that the elastic and plastic parts of the stretching tensor transform as

$$\tilde{D}^{e^*} = H\tilde{D}^e H^T, \quad \tilde{D}^{p^*} = H\tilde{D}^p H^T,$$
$$D^{e^*} = D^e, \qquad D^{p^*} = D^p. \tag{3.38}$$

Equations (3.38) are compatible with the notion that the two forms of the stretching tensors simply correspond to material and spatial representations of the same physical entities. Because the Cauchy stress is unchanged, $T = T^*$, and the stress measure, $\tilde{T}$, also transforms in the same way as U^e.

$$\tilde{T}^* = H\tilde{T}H^T. \tag{3.39}$$

3.2.4 Plastic Yield

Basic to the idea of plasticity is that for given temperature and values of the internal variables there is a limit to the stresses that the material may sustain without undergoing further plastic deformation. This idea is usually expressed by stating that if the stress lies within a certain closed, but possibly evolving, surface in stress space, then the response is purely elastic. Plastic deformation is incipient when the stress first touches the surface. If the material is rate dependent, the rate of plastic stretching increases as the stress moves outside the surface but vanishes if it lies on the surface. If it is rate independent, the stress never lies outside the surface, so plastic deformation can only occur when the stress lies continually on the surface, and the surface must evolve in a compatible way. The yield surface may be expressed as

$$\tilde{f}(\tilde{T}; T, q_n) = 0. \tag{3.40}$$

It is conventional to assume that if $\tilde{T}$ lies in the elastic region within the closed surface for the same values of T and q_n, then $\tilde{f}(\tilde{T}; T, q_n) < 0$. The function $\tilde{f}$ is known as the yield function.

Because $\tilde{T}$ is an objective quantity, that is to say, because it refers to the intermediate reference configuration and therefore is unchanged by rigid body rotations, equation (3.40) is also objective. (It is tacitly assumed that all the internal variables are also objective.) If all the internal variables are scalars, and the material is fully isotropic, then the yield function must be an isotropic function of the stress measure, $\tilde{T}$. When expressed in terms of the Cauchy stress, the yield function becomes $\tilde{f}[(\rho_R/\rho)R^T V^{e-1}TV^{e-1}R; T, q_n]$, but because it is isotropic, the value of $\tilde{f}$ is unaffected by the rotation, R, and so may always be written $\tilde{f}[(\det V^e)V^{e-1}TV^{e-1}; T, q_n]$. Because the material has been assumed to be elastically isotropic with an invertible constitutive relation, the left stretch is an isotropic function of the Cauchy stress, and so the yield function may be written $f(\boldsymbol{T}; T, q_n)$, and the yield surface is given by

$$f(\boldsymbol{T}; T, q_n) = 0. \tag{3.41}$$

The function f must also be objective, and because the Cauchy stress transforms under rigid body rotations as $\boldsymbol{T}^* = \boldsymbol{Q}\boldsymbol{T}\boldsymbol{Q}^T$, f must be an isotropic function with respect to the Cauchy stress. The two isotropic functions, $\tilde{f}$ and f, are distinct, but they have the same values at corresponding values of $\tilde{\boldsymbol{T}}$ and $\boldsymbol{T}$, respectively.

When all the internal variables are scalars, the foregoing is straightforward, but if there is a tensor internal variable, such as a back stress, the picture is more complicated. For example, if one of the internal variables is a tensor, $\tilde{\boldsymbol{A}}$, that transforms under rotations of the initial configuration in the same way as $\tilde{\boldsymbol{T}}$, and if the material is fully isotropic, then the yield function is an isotropic function of two tensor variables. Representations of such functions are far more complicated in general than those with only one tensor variable unless further simplifying assumptions are made. The common assumption known as kinematic hardening assumes that the yield function depends only on the difference of two tensors, say $\tilde{f}[(\rho/\rho_R)\boldsymbol{U}^e\tilde{\boldsymbol{T}}\boldsymbol{U}^e - \tilde{\boldsymbol{A}}; T, q_n]$.

3.2.5 Constitutive Laws for Plastic Flow

To complete the picture, constitutive relations must be given for the elastic and plastic stretching and for the time rates of change of the internal variables. Furthermore, they must be given in such a way that if the histories of stress, temperature, and internal variables are given, then it is possible to construct a corresponding history of the total pure stretching. Conversely, if the history of total stretching is given instead of the history of stress, it must be possible to compute the stress history. These will be called closure relations. Thus, because for any deformation history $\boldsymbol{U}^2 = \boldsymbol{F}^T\boldsymbol{F}$, by differentiation it is found that

$$\frac{d}{dt}(\boldsymbol{U}^2) = 2\boldsymbol{U}^p(\tilde{\boldsymbol{D}}^e + \tilde{\boldsymbol{D}}^p)\boldsymbol{U}^p. \tag{3.42}$$

The elastic stretching, $\tilde{\boldsymbol{D}}^e$, is completely determined by the Gibbs function, as may be seen by reference to Equation (3.34). Because $\boldsymbol{E}^e = \rho_R(\partial g/\partial\tilde{\boldsymbol{T}})$, a short calculation that uses the first of (3.34) and the Maxwell relations in (3.15) shows that

$$\tilde{\boldsymbol{D}}^e = \rho_R\frac{\partial^2 g}{\partial\tilde{\boldsymbol{T}}^2} : \left(\dot{\tilde{\boldsymbol{T}}} - \tilde{\boldsymbol{T}}\frac{\dot{\rho}_R}{\rho_R}\right) + \frac{\partial\boldsymbol{E}^e}{\partial T}\dot{T} - \sum_{n=0}^{N}\frac{\partial Q_n}{\partial\tilde{\boldsymbol{T}}}\dot{q}_n. \tag{3.43}$$

In Equation (3.43) the terms $\rho_R(\partial^2 g/\partial \tilde{\boldsymbol{T}}^2)$ and $(\partial \boldsymbol{E}^e/\partial T)$ are the fourth-order tensor of isothermal elastic compliances and the second-order tensor of thermal expansions, respectively. The last term and the second term in the parentheses will generally represent only a small correction to the thermoelastic terms.

The plastic stretching and the rates of change of the internal variables must be specified by new constitutive relations and must be given in such a way that they are compatible with the entropy inequality (3.21) or (3.22). Generally speaking, it should be expected that each depends on the same set of variables as the Gibbs function:

$$\tilde{\boldsymbol{D}}^p = \tilde{\boldsymbol{\mathcal{D}}}^p(\tilde{\boldsymbol{T}}; T, q_n),$$

$$\dot{q}_n = \tilde{\xi}_n(\tilde{\boldsymbol{T}}; T, q_m), \quad n, m = 0, 1, \ldots, N. \tag{3.44}$$

With the histories of $\tilde{\boldsymbol{T}}$, T, and q_n known, and with $\boldsymbol{U}^p$ regarded as a function of $\tilde{\boldsymbol{T}}$ and $\boldsymbol{U}$ through the decomposition procedures outlined previously, Equation (3.42) may be regarded as a nonlinear ordinary differential equation (ODE) for $\boldsymbol{U}$, requiring only an initial value for $\boldsymbol{U}$. In contrast, if the history of $\boldsymbol{U}$, instead of $\tilde{\boldsymbol{T}}$, is given, then after rearrangement of terms and inversion of the elastic compliances, Equation (3.42) may be regarded as a nonlinear ODE for $\tilde{\boldsymbol{T}}$. Thus the conditions for closure have been met.

A representation for the plastic stretching may be found as follows. As before, if the internal variables are all scalars and if the material is fully isotropic, then the tensor-valued function, $\tilde{\boldsymbol{\mathcal{D}}}^p$, and the rate functions, $\tilde{\xi}_n$, must be isotropic with respect to the stress, $\tilde{\boldsymbol{T}}$. From Equations (3.35), (3.27), and (3.5) the plastic stretching may also be written as

$$\boldsymbol{D}^p = \boldsymbol{V}^{e-1}\boldsymbol{R}\tilde{\boldsymbol{\mathcal{D}}}^p[(\det \boldsymbol{V}^e)\boldsymbol{R}^T \boldsymbol{V}^{e-1}\boldsymbol{T}\boldsymbol{V}^{e-1}\boldsymbol{R}; T, q_n]\boldsymbol{R}^T \boldsymbol{V}^{e-1}$$

$$= \boldsymbol{V}^{e-1}\tilde{\boldsymbol{\mathcal{D}}}^p[(\det \boldsymbol{V}^e)\boldsymbol{V}^{e-1}\boldsymbol{T}\boldsymbol{V}^{e-1}; T, q_n]\boldsymbol{V}^{e-1}, \tag{3.45}$$

where isotropy of the function $\tilde{\boldsymbol{\mathcal{D}}}^p$ has been used. Because the material response is elastically isotropic and invertible, the left stretch is an isotropic function of the Cauchy stress. Therefore, the plastic stretching may be written as

$$\boldsymbol{D}^p = \boldsymbol{\mathcal{D}}^p(\boldsymbol{T}; T, q_n), \tag{3.46}$$

where $\boldsymbol{\mathcal{D}}^p$ is an isotropic function of the stress, $\boldsymbol{T}$. In a completely similar way, if the material is fully isotropic, the rates of change of the scalar internal variables may be written as

$$\dot{q}_n = \xi_n(\boldsymbol{T}; T, q_m), \quad n, m = 0, 1, \ldots, N, \tag{3.47}$$

where all the functions ξ_n are isotropic with respect to the Cauchy stress and therefore depend only on the invariants of $\boldsymbol{T}$. The situation for tensor valued internal variables is somewhat more complicated but will not be discussed here.

From the general representation theorem for isotropic, tensor-valued functions, the most general form of Equation (3.46) may be written (e.g., see Ogden 1984, p. 194) as

$$D^p = f_0 \mathbf{1} + f_1 T + f_2 T^2, \tag{3.48}$$

where f_0, f_1, and f_2 are scalar functions of the invariants of T, as well as of the temperature and the internal variables, and all three functions must vanish when (3.40) or (3.41) is satisfied. Alternatively these three functions may be regarded as functions of the pressure and the two nonzero invariants of the deviatoric stress, S. In fact, there is no loss in generality if T and T^2 are replaced by S and S^2 in (3.48), as well. Recalling Equation (3.20) and that $q_0 = \rho_R$, one finds that another fully general representation of the plastic stretching is

$$D^p = -\frac{\xi_0}{3\rho_R}\mathbf{1} + d_1 S + d_2 \left[S^2 - \frac{1}{3}(\operatorname{tr} S^2)\mathbf{1} \right], \tag{3.49}$$

where ξ_0, d_1, and d_2 are scalar functions of pressure, the invariants of deviatoric stress, temperature, and internal variables, and each function vanishes on the yield surface. The first term in (3.49) accounts for plastic volume change and the last two terms are d^p, which is the deviatoric part of D^p.

The entropy inequality (3.22) now takes the form

$$(\operatorname{tr} S^2)\, d_1 + (\operatorname{tr} S^3)\, d_2 + (p - \rho Q_0)\frac{\xi_0}{\rho_R} - \frac{\rho}{\rho_R} \sum_{n=1}^{N} Q_n \xi_n \geq 0, \tag{3.50}$$

and the alternate inequality (3.23) takes the form

$$(\operatorname{tr} S^2)\, d_1 + (\operatorname{tr} S^3)\, d_2 - \frac{\rho}{\rho_R} \sum_{n=1}^{N} \left(Q_n - \frac{p}{\rho}\frac{\partial \rho_R}{\partial q_n} \right) \xi_n \geq 0. \tag{3.51}$$

In both (3.50) and (3.51) the first two terms represent the rate of deviatoric plastic work, the term with the pressure, p, represents the rate of volumetric plastic work, and the remaining sum of terms involving the Q_n represents the rate of storage of cold work, according to the point of view expressed following Equation (3.1). The two inequalities state that the total rate of plastic work from applied stresses less the rate of storage of cold work cannot be negative. There are no obvious additional restrictions on d_1, d_2, and ξ_n that follow from (3.50) or (3.51), but it is usually assumed that plastic work is always positive. Further, because $\operatorname{tr} S^2 > 0$ during plastic flow, but $\operatorname{tr} S^3$ may be either positive or negative, it seems reasonable to assume that $d_1 > 0$, and that d_2 is odd in $\operatorname{tr} S^3$ so that the second term in (3.50) or (3.51) is always nonnegative.

3.2.6 Flow Potentials

In the theory of viscoplasticity it is often assumed that the rate of plastic flow may be found from a plastic potential, say $\mathcal{P}\,(p, \mathrm{II}_S, \mathrm{III}_S; T, q_n)$, and $\boldsymbol{D}^p = \partial\mathcal{P}/\partial\boldsymbol{T}$, where the invariants of the deviatoric stress are given by

$$\mathrm{II}_S \equiv 1/2\left[(\operatorname{tr}\boldsymbol{S})^2 - \operatorname{tr}\boldsymbol{S}^2\right] = -1/2\operatorname{tr}\boldsymbol{S}^2 = -J_2$$

$$= s_1 s_2 + s_2 s_3 + s_3 s_1 = -1/2\left(s_1^2 + s_2^2 + s_3^2\right),$$

$$\mathrm{III}_S \equiv \det\boldsymbol{S} = 1/3\operatorname{tr}\boldsymbol{S}^3 = J_3$$

$$= s_1 s_2 s_3 = 1/3\left(s_1^3 + s_2^3 + s_3^3\right),$$

(3.52)

and the principal values of $\boldsymbol{S}$ are s_1, s_2, and s_3. The first forms for II_S and III_S are definitions. The second form for II_S follows because $\mathrm{I}_S \equiv \operatorname{tr}\boldsymbol{S} = s_1 + s_2 + s_3 = 0$, and the second form for III_S follows by taking the trace of the Cayley–Hamilton theorem for symmetric tensors, as applied to the deviatoric stress, $\boldsymbol{S}^3 - \mathrm{I}_S\boldsymbol{S}^2 + \mathrm{II}_S\boldsymbol{S} - \mathrm{III}_S\boldsymbol{1} = \boldsymbol{0}$, and noting again that $\boldsymbol{S}$ is traceless.

Rather than II_S and III_S, the notation J_2 and J_3, introduced for the invariants in (3.52), is commonly used in plasticity, and $\sqrt{J_2}$ is often used as a normalizing factor for the components of $\boldsymbol{S}$. With these changes, Equation (3.49) may be rewritten as

$$\boldsymbol{D}^p = -\frac{\xi_0}{3\rho_R}\boldsymbol{1} + \frac{1}{2}\Gamma_1\frac{\boldsymbol{S}}{\sqrt{J_2}} + \frac{1}{3}\Gamma_2\left(\frac{\boldsymbol{S}^2}{J_2} - \frac{2}{3}\boldsymbol{1}\right). \tag{3.53}$$

The response functions ξ_0, Γ_1, and Γ_2 may depend on pressure, the two invariants J_2 and J_3, the temperature, and the scalar internal variables.

If there is a plastic potential, not all of the scalar functions in the two inequalities are independent. First note that because $\boldsymbol{S}$ is deviatoric, by the conventions of linear algebra the gradient of a scalar function of $\boldsymbol{S}$ is also deviatoric. Thus, $\partial\mathrm{II}_S/\partial\boldsymbol{S} = -\boldsymbol{S}$ and $\partial\mathrm{III}_S/\partial\boldsymbol{S} = \boldsymbol{S}^2 - 1/3\,(\operatorname{tr}\boldsymbol{S}^2)\boldsymbol{1}$ from (3.52), and application of the chain rule to the potential yields

$$\boldsymbol{D}^p = \frac{\partial\mathcal{P}}{\partial\boldsymbol{T}} = -\frac{1}{3}\frac{\partial\mathcal{P}}{\partial p}\boldsymbol{1} - \frac{\partial\mathcal{P}}{\partial\mathrm{II}_S}\boldsymbol{S} + \frac{\partial\mathcal{P}}{\partial\mathrm{III}_S}\left[\boldsymbol{S}^2 - \frac{1}{3}(\operatorname{tr}\boldsymbol{S}^2)\boldsymbol{1}\right]. \tag{3.54}$$

A comparison of (3.54) with (3.49) or (3.53) shows that in this case it must be required that $\xi_0 = \rho_R(\partial\mathcal{P}/\partial p)$, $d_1 = 1/2\,\Gamma_1/\sqrt{J_2} = -(\partial\mathcal{P}/\partial\mathrm{II}_S)$, and $d_2 = 1/3\,\Gamma_2/J_2 = (\partial\mathcal{P}/\partial\mathrm{III}_S)$. The spherical and deviatoric parts of the plastic rate would be given by

$$\operatorname{tr}\boldsymbol{D}^p = -\frac{\partial\mathcal{P}}{\partial p},$$

(3.55)

$$\boldsymbol{d}^p = \frac{\partial\mathcal{P}}{\partial\boldsymbol{S}} = \frac{\partial\mathcal{P}}{\partial\boldsymbol{T}} + \frac{1}{3}\frac{\partial\mathcal{P}}{\partial p}\boldsymbol{1}.$$

A less restrictive assumption is that there is a potential for the deviatoric part of the plastic rate, but that the derivative of the potential with respect to pressure need not be related to the rate of plastic volume change. Then only the second equation of (3.55) holds true. Finally, only if $\mathcal{P}$ is independent of p is it generally possible that $\boldsymbol{d}^p = \partial \mathcal{P}/\partial \boldsymbol{T}$.

Although a flow potential is often used in practice for solving problems in viscoplasticity, the existence of such a potential does not seem to have the same theoretical standing as do the fundamental energetic potentials, namely the internal energy, e, the Helmholz free energy, ψ, or the Gibbs energy, g. However, there is some recent work that relates a "principle of maximum dissipation" to the existence of a flow potential (Rajagopal and Srinivasa 1998a, 1998b).

3.3 One-Dimensional Forms

The description of finite deformation plasticity that has been set out in the preceding sections provides a general framework for the examination of plastic instabilities, once the details have been selected that are to be modeled by constitutive equations, yield, and plastic flow. If all features of the full theory were to be implemented, the resulting equations would be virtually impossible to analyze except by numerical methods. For that reason greatly simplified equations that are intended to capture the dominant features of localization have usually been treated in the literature. The simplest example consists of simple shearing in one dimension. In this case, it is assumed that there may be rapid variations in all field quantities in a direction normal to the plane of a shear band, but that there are no variations (or only very small or slow variations) within the plane of the band itself. By comparing the fully general energy and momentum equations with the usual one-dimensional equations, it is easy to see what specializations or approximations have to be made in order that the one-dimensional version should have some validity.

3.3.1 Specializations and Approximations

The general balance of momentum is given by the second equation of (3.13), repeated here without body force, and the one-dimensional specialization is given below it.

$$\operatorname{div} \boldsymbol{T} = \rho \dot{\boldsymbol{v}},$$
$$s_y = \rho \dot{v}. \tag{3.56}$$

In the second equation of (3.56) s is the shear stress acting on planes $y = \text{const}$, and v is the particle velocity parallel to those planes and in the same direction as the shear stress. The stress normal to the plane of shearing and the shear tractions perpendicular to the direction of shearing must both be essentially constant in the y direction to satisfy the other two components of (3.56).

Similarly, the general balance of energy, repeated from (3.25), and the one-dimensional specialization follow:

$$\rho c_E \dot{T} = -\text{div}\, \boldsymbol{q} + \frac{\rho}{\rho_R} T \frac{\partial \tilde{\boldsymbol{T}}}{\partial T} : \dot{\boldsymbol{E}}^e + \boldsymbol{T}:\boldsymbol{D}^p - \frac{\rho}{\rho_R} \sum_{n=0}^{N} \left(Q_n - T \frac{\partial Q_n}{\partial T} \right) \dot{q}_n,$$

$$\rho c \dot{T} = k T_{yy} + \beta s \dot{\gamma}^p.$$

$$(3.57)$$

Here it is clear that there are several significant approximations, which are listed as follows.

a. The specific heat at constant strain has been replaced by a generic specific heat, $c_E \to c$ (meaning c_E is replaced by c). For metals this is a modest approximation because the difference between the specific heats at constant stress and constant strain is typically less than a few percent of either one. Furthermore, the specific heat is usually taken to be constant, or at most a function of temperature (see Table 3.1).
b. The heat flux is given by Fourier's law in one dimension. If the thermal conductivity is taken to depend on temperature, then the term becomes $\partial[k(T)T_y]/\partial y$.
c. The thermoelastic term is ignored. For an elastically isotropic material it is easily shown that

$$\frac{\rho}{\rho_R} T \frac{\partial \tilde{\boldsymbol{T}}}{\partial T} : \dot{\boldsymbol{E}}^e = T \frac{\partial \boldsymbol{T}}{\partial T} : \frac{1}{2} \left(\dot{\boldsymbol{V}}^e \boldsymbol{V}^{e^{-1}} + \boldsymbol{V}^{e^{-1}} \dot{\boldsymbol{V}}^e \right) \approx -3K\alpha\, T \text{tr}\left(\dot{\boldsymbol{V}}^e \boldsymbol{V}^{e^{-1}} \right).$$

$$(3.58)$$

In (3.58) K is the bulk modulus, α is the coefficient of thermal expansion, and $\text{tr}(\dot{\boldsymbol{V}}^e \boldsymbol{V}^{e^{-1}})$ is the rate of elastic volumetric expansion. The last expression is reached by using the approximation $\partial \boldsymbol{T}/\partial T \approx -3K\alpha \boldsymbol{1}$, which ignores the effect of small nonvolumetric elastic distortions.
d. The plastic work, $\boldsymbol{T}:\boldsymbol{D}^p$, is replaced by the simple expression $s\dot{\gamma}^p$. In a one-dimensional motion such as $x = X + u(y,t)$, $y = Y$, and $z = Z$ where y is the coordinate normal to the plane of the shear band, and $u(y,t)$ is a displacement parallel to the plane of the band, the total stretching, $\boldsymbol{D}$, has only two nonzero entries, $2D_{12} = 2D_{21} = v_y$, where v is the velocity, $v = \dot{u}$. If the deviatoric Cauchy stress also has only two nonzero components, $S_{12} = S_{21} = s$, then $\boldsymbol{S}^2$ is diagonal, and it is easily checked that $\text{II}_S = -s^2$, and $\text{III}_S = 0$. With plastic incompressibility assumed as well, then for a fully

Table 3.1. *Difference between specific heats at zero pressure and 300 K*

Material	Density, ρ (10^3 kg/m^3)	Young's modulus, E (GPa)	Shear modulus, μ (GPa)	Poisson's ratio, v	Thermal expansion, α (μm/m K)	Specific heat, c (J/kg K)	$c_{\bar{T}} - c_E$ (J/kg K)	$\Delta c/c$ %
Al	2.7	69	26	0.33	23.2	903	36	4
Be	1.85	300	140		11.5	1825	23	1
Cu	8.96	120	46	0.355	16.8	385	12	3
Mo	9.01	330	130		5	248	2	<1
Ni	8.90	200	80		12.7	444	7	1.5
Ta	16.6	180	70		6.5	141	1	<1
Ti	4.54	110	40		8.5	522	6	1
W	19.3	380	150		4.5	133	1	<1
U	18.9	170	72		13.5	116	2	2
Brass	8.4	110	41	0.331	21	380	15	4
Steel	7.8	200	81	0.29	11.5	460	7	1.5

Note: The first six columns of data were taken from Bai and Dodd (1992). The difference of specific heats was calculated from Equation (2.39) in the form appropriate for isotropic materials, i.e., $\Delta c = (3\alpha/\rho)\,[p + (E\alpha T)/(1 - 2v)]$.

isotropic material, Equation (3.53) reduces to

$$D^p = \frac{1}{2}\Gamma\left(|s|, T, q_n\right)\frac{S}{|s|} \tag{3.59}$$

provided that $\Gamma_2 = 0$ when $\mathrm{III}_S = 0$. This last assumption is reasonable because according to (3.50) and (3.52), d_2 (or Γ_2) then has the desirable property that its contribution to the entropy inequality is always nonnegative if d_2 is odd in III_S and vanishes when $\mathrm{III}_S = 0$, as was observed previously. Thus the plastic stretching also has only two nonzero components, which are identified as $\dot{\gamma}^p = 2D^p_{12} = 2D^p_{21}$. Under the conditions described, the plastic work is given exactly by $T\!:\!D^p = s\dot{\gamma}^p = |s|\,\Gamma$, so it will be an approximation for nearby stress states as well. The linear elastic approximation for the elastic stretching also has only two nonzero components, $\dot{s}/\mu = 2D^e_{12} = 2D^e_{21}$, where μ is the elastic shear modulus. As a consequence of these considerations, the approximate relationship among the elastic, plastic, and total stretching is the same as in small strain theory:

$$v_y = \dot{s}/\mu + \dot{\gamma}^p. \tag{3.60}$$

e. In the first equation of (3.57) the last terms in parentheses are thermoplastic terms that arise from the changes in energy and entropy caused by changing internal variables only. To understand the essence of their meaning, observe that if the Helmholz free energy could be partitioned into two terms, $\psi = \hat{\psi}(E^e, T) + \phi(q_n)$, then the first term represents only thermoelastic energy, and the second one represents the stored energy of cold work, which depends only on the internal variables. It seems unlikely that the total energy would decompose in this fashion, but if it did, all the Q_n would be independent of temperature, and the remaining sum would give $(\rho/\rho_R)\sum_{n=0}^{N}[Q_n - T(\partial Q_n/\partial T)]\dot{q}_n = \rho\dot{\phi}$. In any case, all of the effect of the changing internal variables and their contribution to the total energy is contained in the coefficient β, which in the second equation of (3.57) is taken to be constant and slightly less than 1.

In stating the last approximation, most authors cite the experimental work of Farren and Taylor (1925) and Taylor and Quinney (1937), who seem to have been the first to attempt to measure the amount of plastic work that is converted to heat during plastic deformation. They concluded that approximately 85% to 95% of the plastic work is converted to heat, which is equivalent to asserting that the final sum of terms in the first equation of (3.57) is roughly proportional to the plastic work and that the coefficient of proportionality is in the range 0.05–0.15. It has been known in the metallurgical literature for a long time that the so-called "stored energy of cold work" can vary enormously depending on the material, temperature, and extent of deformation. This work has been summarized by Bever, Holt, and Titchener (1972). Recently there have been both theoretical and experimental studies that attempt to clarify and model this phenomenon (e.g., see Aravas, Kim, and Leckie 1990; Kamlah and Haupt 1997; Bodner and Lindenfeld 1995; Hodowany, Ravichandran, Rosakis, and Rosakis 2000; and Rosakis, Rosakis, Ravichandran, and Hodowany 2000). From the form of (3.25) or the first equation of (3.57) it is apparent that there is a lot of scope for modeling of the physical processes that generate the internal variables and the manner in which they contribute to the total energy.

3.3.2 Isotropic Work Hardening and the Stored Energy of Cold Work: An Example

As one of the simplest possible phenomenological examples, consider a fully isotropic material with one positive internal variable, κ, that characterizes the extent of work hardening. Such a simple model may be appropriate for limited plastic strains in a relatively pure, polycrystalline metal in which plastic slip alone accounts for the hardening with no secondary mechanisms such as twins, precipitates, or other phases. Suppose that the yield function in (3.41) may be written $f(|s|, T, \kappa) = \hat{f}(|s|, T) - \kappa$, which has the property

$$\hat{f}(|s|, T) \begin{cases} < \kappa, & \text{elastic} \\ = \kappa, & \text{onset of yield.} \\ > \kappa, & \text{plastic flow} \end{cases} \tag{3.61}$$

Because there is only one internal variable, there is also only one internal thermodynamic force, $Q = \rho_R (\partial \psi / \partial \kappa)$, and because the motion has been assumed to be one-dimensional shearing as noted in the first approximation above, the density, ρ, is constant. Because the internal and Helmholz energies must increase as κ increases because of an increasing number of dislocations, it may be assumed that $Q \geq 0$.

The momentum equation remains the same as the second equation of (3.56), but the energy equation does not yet look like the second equation of (3.57). At this stage the equations read

$$\rho \dot{v} = s_y,$$

$$\rho c \dot{T} = (kT_y)_y + s\dot{\gamma}^p - \left(Q - T \frac{\partial Q}{\partial T} \right) \dot{\kappa}. \tag{3.62}$$

For the system to be completed, constitutive laws must be given for the three functions Q, $\dot{\gamma}^p$, and $\dot{\kappa}$:

$$Q = \hat{Q}(|s|, T, \kappa),$$

$$\dot{\gamma}^p = \Gamma(|s|, T, \kappa)\frac{s}{|s|}, \tag{3.63}$$

$$\dot{\kappa} = \xi(|s|, T, \kappa).$$

These constitutive functions cannot be completely arbitrary; they are subject to a number of restrictions.

Because no plastic deformation can occur in the elastic region, but by definition always occurs outside it, the constitutive function for the plastic strain rate must have the property

$$\Gamma(|s|, T, \kappa) \begin{cases} = 0 & \text{if } \hat{f}(|s|, T) \leq \kappa, \\ > 0 & \text{if } \hat{f}(|s|, T) > \kappa. \end{cases} \tag{3.64}$$

Furthermore, the reduced entropy inequality now reads

$$|s|\Gamma(|s|, T, \kappa) - \hat{Q}(|s|, T, \kappa)\xi(|s|, T, \kappa) \geq 0. \tag{3.65}$$

Finally, where $\hat{Q} > 0$, the rate of change of the work hardening parameter need only satisfy

$$\xi(|s|, T, \kappa) \leq \frac{|s|\,\Gamma(|s|, T, \kappa)}{\hat{Q}(|s|, T, \kappa)}. \tag{3.66}$$

Note that Equation (3.66) only limits the magnitude of the rate of work hardening, but it does not require that the rate should actually be positive. In fact, if the material is in the elastic range, $\hat{f}(|s|, T) < \kappa$, it demands that $\xi \leq 0$, because the numerator in (3.66) is zero in that case. Furthermore, it is not inconsistent to have $\xi(|s|, T, \kappa) < 0$ even when the material is in the plastic range with $|s|\Gamma(|s|, T, \kappa) > 0$. Thus, even the simplest one-dimensional model has scope for elementary modeling of both static and dynamic recovery and a consequent decrease of the work hardening parameter. This fact has been pointed out by Anand (1985) and by Anand and Brown (1987).

In particular, it must be noted that there need be no exact correlation between changes in the internal variable and changes in the isochoric part of plastic strain. This feature is completely contrary to the definition, apparently first given by Rice (1971), and subsequently used by many authors since, that increments of plastic strain occur as a consequence of increments in the internal variables. In the present notation that definition may be written as $(\delta \mathbf{E})^p \equiv \sum_{n=1}^{N} (\partial \mathbf{E}/\partial q_n)\delta q_n$. It must be emphasized again that in the present view, plastic strain may be calculated as a consequence of deformation, but except possibly for plastic volume change, it is never to be regarded as an internal variable itself in any fundamental way or even directly connected to internal variables, as suggested by Rice's definition. Indeed, the plastic strain and the internal variables must be calculated independently as suggested by two limiting cases. First, note that the internal variables may all saturate at extreme deformations and elevated temperatures, but plastic straining continues. In this case, which may be called saturation, $\delta q_n = 0$, but $(\delta E)^p \neq 0$. At the other extreme, the stress state may lie well within the yield surface so that no plastic deformation can occur, but thermal excursions may alter the internal variables. In this case, which may be called static recovery, $\delta q_n \neq 0$, but $(\delta E)^p = 0$. In dynamic recovery stored energy corresponding to the internal variables may sometimes be released during plastic deformation if the temperature is high enough. If the theory is to include saturation and recovery, there can be no direct relationship between increments of internal variables and increments of plastic strain.

If the energy equation is to reduce to the second equation of (3.57), and if the term $T(\partial Q/\partial T)\dot{\kappa}$ is negligible, then the coefficient, β, could be interpreted as $\beta = 1 - [\hat{Q}(|s|, T, \kappa)\xi(|s|, T, \kappa)/|s|\Gamma(|s|, T, \kappa)]$, which corresponds to the relationship $s\dot{\gamma}^p - Q\dot{\kappa} = \beta s\dot{\gamma}^p$. More generally, if the thermoplastic term is not negligible, the coefficient could be interpreted as

$$\beta = 1 + \frac{T^2(\partial/\partial T)\{[\hat{Q}(|s|, T, \kappa)]/T\} \cdot \xi(|s|, T, \kappa)}{|s|\Gamma(|s|, T, \kappa)}. \tag{3.67}$$

In general it would appear to be unlikely for β to be constant, given the nature of the other constitutive functions, except in cases in which work hardening saturates, that is, where $\xi \to 0$, and $\beta \to 1$. If significant dynamic recovery occurs with rapid release of energy previously stored as cold work, it would appear that β in (3.67) could even be greater than one. This would occur, for example, if $\hat{Q}/T$ is a decreasing function of T while ξ is negative.

Recently an attempt has been made to measure β directly in a series of Hopkinson bar compression experiments on Al 2024-T3 and α-Ti (Hodowany et al. 2000). With rapid homogeneous loading at rates of the order of 10^3 s^{-1}, the temperature change in the specimen was measured, as well as the usual stress and strain variables. Then because heat conduction contributes nothing to the balance of energy (except in a negligible region near the contacts between the specimen and the loading bars), the coefficient β may be found from the expression

$$\beta = \rho c\dot{T}/\dot{W}^p, \tag{3.68}$$

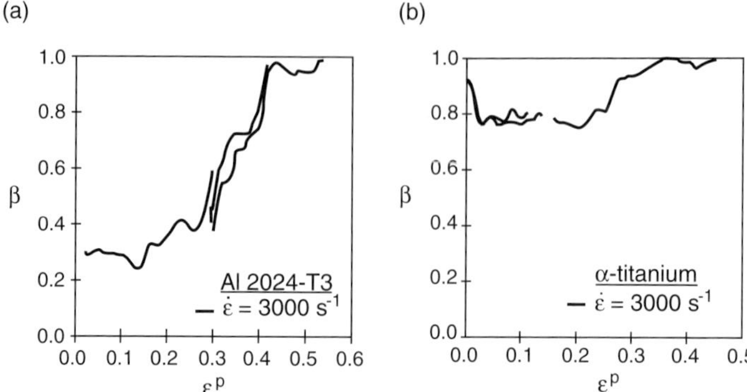

Figure 3.6. Experimental measurement of the fraction of plastic work that is converted into heat in an Al alloy (panel a) and alpha titanium (panel b). Neither flow stress nor fraction of work converted to heat were rate dependent in the Al alloy, but both were rate dependent in the titanium. (Reproduced from Hodowany et al. 2000). Copyright 2000 by Sage Publications.)

where the denominator $\dot{W}^p$ is the rate of plastic work, as calculated from data obtained in the experiment. Although there are several experimental uncertainties, particularly near the start of the experimental record as loading commences, it is clearly evident from plots of β versus plastic strain (see Figure 3.6) that β is much less than one initially as plastic deformation and work hardening begin, but that it approaches one asymptotically with increasing plastic deformation.

4

Models for thermoviscoplasticity

From the continuum point of view, as presented in Chapter 3, even the simplest version of rate-dependent plasticity requires the specification of three constitutive equations, as noted in Equation (3.63), in addition to a quasi-static yield surface. More would be required if more than one scalar internal variable were used to model the internal state of the material or if the function Γ_2 in (3.53) were not assumed to vanish identically. In the literature of adiabatic shearing, the usual practice is to specify a flow rule corresponding to the second equation of (3.63) and a work hardening relation corresponding to the third equation of (3.63), but to ignore the specification of the thermodynamic force, $Q = \hat{Q}(s, T, \kappa)$, which is conjugate to the internal variable, κ. Instead, the coefficient β in the second equation of (3.57) is given as a constant, which in effect defines $\hat{Q}$ indirectly (within a function of s and κ) through integration of Equation (3.67). The resulting theory may be called a J_2-flow theory of thermoviscoplasticity, and it may be represented by the equations

$$\boldsymbol{D}^p = \frac{1}{2}\Gamma(\sqrt{J_2}, T, \kappa)\frac{\boldsymbol{S}}{\sqrt{J_2}},$$

$$\dot{\kappa} = \xi(\sqrt{J_2}, T, \kappa),$$
(4.1)

where Γ vanishes inside or on the yield surface and is positive outside it, as in (3.64). In the case of one-dimensional shearing, $\sqrt{J_2} = |s|$, and (4.1) reduces to (3.59) and the third of (3.63). It is often useful to define the equivalent plastic strain and strain rate as

$$\dot{\gamma}^p_{\text{eq}} \equiv \sqrt{2\boldsymbol{D}^p : \boldsymbol{D}^p} = \Gamma,$$

$$\gamma^p_{\text{eq}} = \gamma^p_{\text{eq}_0} + \int_{t_0}^{t} \dot{\gamma}^p_{\text{eq}}\, dt.$$
(4.2)

77

From (4.1) and (4.2) it follows that the plastic work may be expressed as

$$\dot{W}^p \equiv \boldsymbol{S} : \boldsymbol{D}^p = \sqrt{1/2\boldsymbol{S}:\boldsymbol{S}}\sqrt{2\boldsymbol{D}^p:\boldsymbol{D}^p} = \sqrt{J_2}\,\Gamma = |s|\dot{\gamma}^p_{\mathrm{eq}}, \qquad (4.3)$$

where the last form is appropriate for the one-dimensional case.

Theories in which the yield surface depends on stress through both J_2 and J_3 should logically include the second flow function Γ_2 in Equation (3.53) or (3.54), as well as Γ_1. For example, an isotropic flow function that depends only on J_2 (for a given temperature and fixed values of the scalar internal variables, tensor variables being excluded) must of necessity be a von Mises type of material because first yield can only occur on a surface in stress space given by $J_2 =$ const. In particular, a Tresca type of yield surface, which cannot be expressed by $J_2 =$ const, is excluded. Theories of a more general type have been little developed to the author's knowledge. Only J_2-flow theories will be considered from here on.

Once the choice has been made that β is to be constant, the qualitative nature of solutions to the simple one-dimensional equations in many cases is unaffected by its exact value, because often it can be absorbed into a scaled variable in such a way that it no longer appears in the nondimensional equations. In recovering the solution in dimensional form, the value of β does matter, of course, but for many analytical purposes it may simply be set equal to one. This point will be expanded further in the next and subsequent chapters, but for now it will be understood that β is a constant that is equal to or only slightly less than one.

The remainder of this chapter describes some typical constitutive equations that have been used to describe work hardening and plastic flow with J_2-flow theory.

4.1 Work Hardening

It has been assumed that Equation (3.61) describes the limit of the elastic region. This equation states that for a given, fixed value of the work hardening parameter, there is a curve in shear stress-temperature space that bounds the elastic region. The situation envisioned is shown schematically in Figure 4.1. Note that the curve must be symmetric in the shear coordinate, as required by isotropy. To be more concrete, one choice for the work hardening parameter is to let κ be the quasi-static yield stress in shear at a reference temperature, T_0. This choice for the internal variable simply parameterizes the curves $\hat{f}(|s|, T) =$ const according to the intersection of the curve with the stress axis at $T = T_0$, and then normalizes the curves so that $\hat{f}(\kappa, T_0) = \kappa$. Implicit in this parameterization is

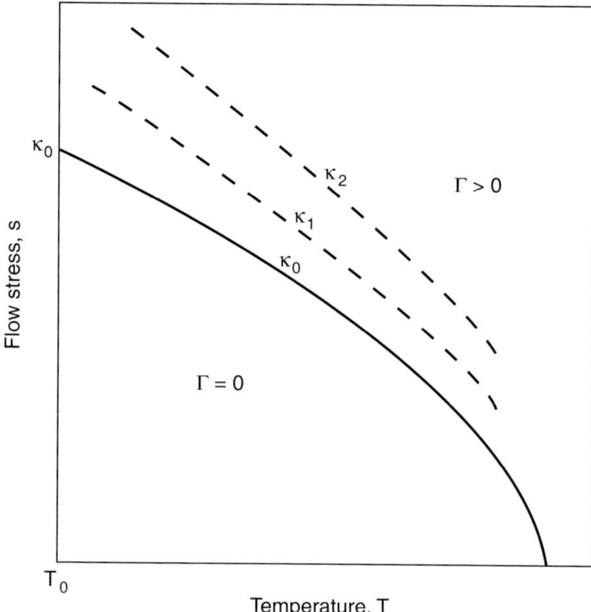

Figure 4.1. The yield surface in stress-temperature space for a sequence of fixed values of work hardening. The curves, $\hat{f}(s, T) = \kappa$, for various fixed values of κ, are normalized so that $\kappa = s$ at $T = T_0$. Plastic flow only occurs when the point (s, T) lies above the current curve.

the assumption that the isothermal yield stress increases with increasing plastic deformation so that as κ increases the elastic region expands. This idea is also shown schematically in Figure 4.1. Other parameterizations are possible, of course, and various authors have proposed many variations.

Work hardening refers to the idea that the yield stress of a material generally tends to increase with plastic deformation although if the temperature is high enough, recovery processes may be initiated. This idea is shown schematically in Figure 4.2. Besides the yield surface where $\Gamma(|s|, T, \kappa) = 0$ and plastic flow ceases, another structural surface where work hardening vanishes may be defined by $\dot{\kappa} = \xi(|s|, T, \kappa) = 0$. This curve may coincide with the yield surface over all or only over part of its length, as shown in the figure. Work hardening occurs in regions where $\xi > 0$, and softening where $\xi < 0$. If there is a region in which $\xi < 0$, but $\Gamma = 0$, as in the right side of Figure 4.2, then the yield surface must be moving inward toward the origin, but no plastic deformation can occur until the load point actually touches and moves outside the yield surface. The possibility of independent structural surfaces was noted by Scheidler and Wright (2001), but the idea was not developed further. It seems clear, however,

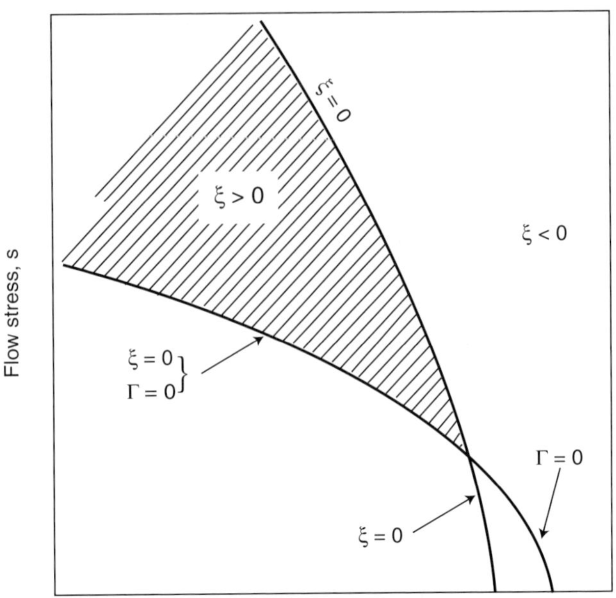

Figure 4.2. Structural surfaces in stress-temperature space, where $\dot{\kappa} = \xi(s, T, \kappa) = 0$ for a given value of κ, need not coincide with the yield surface. The structural parameter increases (decreases) in the region where $\xi > 0 (\xi < 0)$.

that the added feature of independent structural surfaces leaves room for future modeling of complex thermal behavior of materials.

Measurement of isothermal, quasi-static, stress–strain curves places further restrictions on the constitutive functions $\xi(|s|, T, \kappa)$ and $\Gamma(|s|, T, \kappa)$, which were introduced in Chapter 3, and which determine the evolution of the work hardening parameter and the plastic strain. These functions must be consistent with quasi-static work hardening in the limit that the yield surface is approached from the outside. Therefore, because we may write

$$d\kappa = \frac{d\kappa}{d\gamma^{p}_{eq}} d\gamma^{p}_{eq} = \frac{\dot{\kappa}}{\dot{\gamma}^{p}_{eq}} d\gamma^{p}_{eq} = \frac{\xi(|s|, T, \kappa)}{\Gamma(|s|, T, \kappa)} d\gamma^{p}_{eq} \qquad (4.4)$$

for slow monotonic loading, in order to construct the quasi-static stress–strain curve, we need to know the limit of ξ/Γ as $\kappa \to \hat{f}(|s|, T)$ and $\Gamma \to 0$, say $\lim_{\Gamma \to 0}(\xi/\Gamma) = \hat{h}(s, T)$. For the present it will be supposed that the temperature is low enough that ξ vanishes on the yield surface and that $\hat{h}$ is finite.

4.1.1 Work Hardening Without History Effects

The very simplest case and the one most often used in viscoplasticity is equivalent to the assumption that $\xi(|s|, T, \kappa) = \phi(\kappa)\Gamma(|s|, T, \kappa)$. Then (4.4) reduces to $d\kappa = \phi(\kappa)\, d\gamma_{eq}^{p}$, which indicates an equivalence between κ and γ_{eq}^{p}. Upon integration and inversion the relation may be written $\kappa = h(\gamma_{eq}^{p})$. Then the first equation of (4.2) becomes $\dot{\gamma}_{eq}^{p} = \Gamma(|s|, T, \gamma_{eq}^{p})$, or upon solving for the stress, $|s| = f(\gamma_{eq}^{p}, \dot{\gamma}_{eq}^{p}, T)$. Although there is a certain plausibility to this last relationship as a generalization to an empirical, quasi-static relation between stress and plastic strain, it must be regarded only as a very special case. Furthermore, it cannot generally reproduce the so-called history effects described in Chapter 1.

4.1.2 Quasi-Static, Isothermal Stress–Strain Curves

To illustrate how the function ξ and the work hardening response are related, consider the case of slow isothermal changes, which must occur essentially on the yield surface. Because isothermal increments of κ and s on the yield surface are related by $\delta\kappa = \{[\partial \hat{f}(s, T)]/\partial s\}\delta s$, from the limit of (4.4) where $\delta\kappa = \hat{h}(s, T)\delta\gamma_{eq}^{p}$ isothermal increments of stress must be related to increments of plastic strain by $\hat{f}_{s}\delta s = \hat{h}\delta\gamma_{eq}^{p}$ or

$$\frac{\delta s}{\delta\gamma_{eq}^{p}} = \frac{\hat{h}(s, T)}{\hat{f}_{s}(s, T)}. \tag{4.5}$$

The plastic slope of the isothermal stress–strain curve, that is, the left-hand side of (4.5), expressed as a function of stress and temperature, may be determined experimentally. Each integration of the equation at a distinct fixed temperature results in a different relation between stress and plastic strain.

As an example, isothermal work hardening in a quasi-static test and at a given reference temperature can often be described empirically by a power law relation

$$s = \kappa_0 \left(1 + \frac{\gamma^p}{\gamma_0}\right)^n = K_0 \left(\gamma_{eq}^p\right)^n. \tag{4.6}$$

In (4.6), γ^p is the plastic shear strain, and κ_0, γ_0, and n are simply fitting parameters. Here κ_0 is the initial yield stress, n is the work hardening exponent (generally much less than one), and γ_0 is a characteristic strain that may be used to improve the fit. The second form of the equation is just the obvious rewriting with $K_0 = \kappa_0/(\gamma_0)^n$ and $\gamma_{eq}^p = \gamma_0 + \gamma^p$. If (4.6) is solved for the plastic strain, which is then added to the elastic strain, s/μ, the resulting expression

$$\gamma = \frac{s}{\mu} + \left(\frac{s}{K_0}\right)^{1/n} \tag{4.7}$$

is just the Ramberg–Osgood equation, which has been widely used.

Although plastic strain is easily determined in a simple tension or torsion test, it is not a fundamental parameter that is suitable for use in the general theory as a state variable. However, it is still necessary to make a connection between the evolution of the work hardening parameter and standard laboratory measurements of stress–strain curves. This is not always a straightforward process, and it is probably best done by appealing to modeling of fundamental physical processes. It may also be done, at least tentatively, in an incremental and self-consistent fashion. The following argument will serve as illustration.

First differentiate (4.6), and then eliminate the plastic strain.

$$ds = \frac{n}{\gamma_0}\left(\frac{s}{\kappa_0}\right)^{-1/n} s\,d\gamma^P \tag{4.8}$$

At the reference temperature T_0 it will be recalled that $s = \kappa$, and therefore, at $T = T_0$ (4.8) is equivalent to

$$d\kappa = \frac{n}{\gamma_0}\left(\frac{\kappa}{\kappa_0}\right)^{-1/n} dW^P, \tag{4.9}$$

where dW^P is the increment of plastic work that induces the increment in the work hardening parameter, $d\kappa$. Selecting dW^P to replace $s\,d\gamma^P$ effectively eliminates plastic strain as a primary variable and replaces it with a quantity that can be specified by constitutive functions and that also has meaning in three-dimensional deformations.

Equation (4.9) has been obtained as an incremental approximation to a quasi-static stress–strain curve at the limiting temperature, T_0, but now suppose that it applies at all temperatures and strain rates. That is to say, suppose that

$$\xi(s, T, \kappa) = \frac{n}{\gamma_0}\left(\frac{\kappa}{\kappa_0}\right)^{-1/n} s\Gamma(s, T, \kappa). \tag{4.10}$$

Equation (4.10) is obviously only one of many generalizations that could be made from the quasi-static case, but it has the virtue of being simple and direct, it has the correct limiting behavior, and it depends on proper state variables but does not depend explicitly on plastic strain.

The initial yield surface for one-dimensional shearing can always be expressed as $s = \kappa_0 g(T)$ where $g(T)$ is most commonly a decreasing function of temperature, and at the reference temperature, $g(T_0) = 1$. The number κ_0 is the initial yield stress at the reference temperature. Now assume that the quasi-static yield function always has the same general form, which is

$$s = \kappa g(T).$$

(4.11)

Equation (4.11) assumes that the family of quasi-static yield surfaces is self-similar in a particular way. Again this is a simple and direct generalization, but others are certainly possible.

As $\kappa \to s/g(T)$ from the outside of the yield surface, Equation (4.10) shows that the limiting value of the ratio ξ/Γ is just the expression

$$\hat{h}(s, T) = \frac{n}{\gamma_0} \left[\frac{s/g(T)}{\kappa_0} \right]^{-1/n} s.$$

(4.12)

Now according to (4.5) because $\kappa = s/g(T)$ and therefore $\hat{f}_s = g^{-1}$, the isothermal, quasi-static, stress–strain curve at an arbitrary temperature can be calculated from the differential relationship

$$\frac{\delta s}{\delta \gamma^p} = g(T) \frac{n}{\gamma_0} \left[\frac{s}{\kappa_0 g(T)} \right]^{-1/n} s.$$

(4.13)

With the recognition that the initial yield stress at the fixed temperature, T, must be $s_0 = \kappa_0 g(T)$, Equation (4.13) can be integrated to give the quasi-static, isothermal stress–strain curve at an arbitrary temperature

$$s = \kappa_0 g(T) \left[1 + g(T) \frac{\gamma^p}{\gamma_0} \right]^n.$$

(4.14)

At the reference temperature T_0, Equation (4.14) reduces to (4.6), as it should, but at higher temperatures, $T > T_0$, the material is softer in two ways. Not only is the initial yield stress reduced to $\kappa_0 g(T)$, but in addition the effective characteristic strain has been increased to $\gamma_0/g(T)$, which has the effect of reducing the rate of hardening at a given plastic strain. Although the example as given may be somewhat artificial, it does mimic the frequently observed behavior that at higher temperatures the stress–strain curve is flatter than would be obtained by simply reducing the whole curve by a common factor. Thus, although the yield surface is assumed to be self-similar, as pointed out above, the isothermal stress–strain curves are not. The cumulative result is shown in Equation (4.14). At temperatures lower than the reference temperature, the stress–strain curve is affected in a manner that is the reverse of the preceding description. Figure 4.3 illustrates the behavior.

A generalization of the procedure above shows that this result is completely typical. Thus, if the quasi-static stress–strain curve at the reference temperature, T_0, is given by a monotonic function, $s = h(\gamma^p)$, then the slope

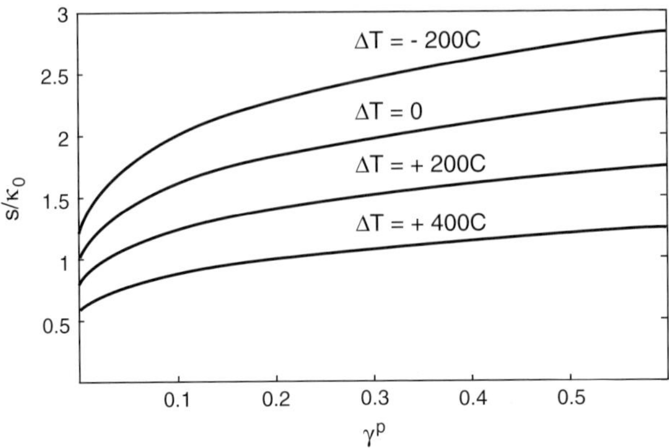

Figure 4.3. Quasi-static, isothermal stress–strain curves plotted from Equation (4.14) for linear thermal softening, with $\bar{a} = 0.001\,°\text{C}^{-1}$, $\gamma_0 = 0.01$, and $n = 0.2$. Note that the ratio between the top and bottom curves is $2:1$ at $\gamma_p = 0$, but more than $2.2:1$ at $\gamma_p = 0.6$.

of the curve would be $\delta s/\delta\gamma^P = h'[h^{-1}(s)]$, and the assumed rate of work hardening, following a procedure similar to that leading to (4.10), would be $\xi(s, T, \kappa) = h'[h^{-1}(\kappa)]\kappa^{-1}s\Gamma(s, T, \kappa)$. Then because of the assumed self-similarity of the family of yield surfaces, the quasi-static, isothermal response at an arbitrary temperature will be predicted to be

$$s = g(T)h[g(T)\gamma^P].\tag{4.15}$$

This effect is a direct consequence of the assumptions that increments of work hardening are proportional to increments in plastic work and that the family of yield surfaces is self-similar. Because of thermal softening the stress at a given plastic strain at an elevated temperature is less than at the reference temperature, which reduces the increment of plastic work and hence the increment in work hardening for a given increment of plastic strain.

It must be emphasized that (4.14) and (4.15) would apply only for isothermal response and would not hold for arbitrary thermal histories. In the general case the response must be found from the incremental forms of the constitutive laws.

4.1.3 *Thermodynamic Consistency of Cold Work and Work Hardening*

To be thermodynamically consistent, Equation (3.65) requires that $1 - (\hat{Q}\xi/|s|\Gamma) \geq 0$, which in the present case reduces to

$$1 - \hat{Q}(|s|, T, \kappa)h'[h^{-1}(\kappa)]\kappa^{-1} \geq 0.\tag{4.16}$$

In the special case that equivalent plastic strain may be used as a work hardening parameter and that work hardening may be expressed as a power law as in (4.6), the ratio h'/κ reduces to $[nK(\gamma_{eq}^{p})^{n-1}]/[K(\gamma_{eq}^{p})^{n}] = n/\gamma_{eq}^{p}$. Because $\hat{Q} > 0$, to avoid violating (4.16) as the equivalent plastic strain vanishes, the thermodynamic force must satisfy

$$\hat{Q} = \hat{Q}_0 \frac{\gamma_{eq}^{p}}{n} + O\left(\gamma_{eq}^{p}\right)^2 \quad \text{with } 0 \le \hat{Q}_0 < 1. \tag{4.17}$$

Thus, the rate of storage of cold work at small plastic strains and large rates of work hardening must itself be small. Equation (4.16) again illustrates the intimate connection between the rate of work hardening and the rate of storage of cold work as internal energy. In general, when work hardening is strong and $h'(\cdot)$ is large, the rate of storage of cold work must be small to avoid violation of (4.16).

The considerations in this section have been largely empirical, but they illustrate the constructive procedures that may be used to obtain thermodynamically consistent constitutive functions that do not depend on plastic strain as an internal variable, even though plastic strain must inevitably appear in the raw laboratory data. Forms other than those used above for the constitutive functions may be suggested by theoretical or other considerations, as will be illustrated in the following section, but it always appears possible to develop those functions in terms of an internal variable that does not depend on plastic strain.

4.2 Plastic Flow: Simple Phenomenological and Physical Models

Analytical representations for viscoplastic flow have been given by many authors. Most have made no distinction between the constitutive function for the rate of work hardening, denoted as $\xi(s, T, \kappa)$ in this book, and the rate of plastic flow, denoted here as $\Gamma(s, T, \kappa)$. Rather it has been common to combine both into a single function as a so-called dynamic flow stress, $s = \hat{s}(\gamma^{p}, \dot{\gamma}^{p}, T)$, as was done originally by Bai (1982) and Clifton (1980) in their pioneering examinations of adiabatic shearing. As was shown in the last section, such a representation of the flow stress is consistent with the notion of an internal variable for work hardening, provided that the rate of work hardening may be written as $\dot{\kappa} = \xi(|s|, T, \kappa) = \phi(\kappa)\Gamma(|s|, T, \kappa)$. As was also noted previously, such a representation cannot capture "history effects" in high-rate jump tests.

Virtually all authors use some form of empirical, power law for work hardening, and often the elastic strain is completely eliminated for consideration of adiabatic shear bands. This last approximation can be a valuable simplification for analysis without introducing significant error, as will be shown in the next

chapter. It usually is not difficult to convert the power law for work hardening in terms of plastic strain into a representation that uses a work hardening
parameter, as has been done in the last section.

Mathematical models for the effect of temperature and strain rate generally
fall into one of two categories. Either the modeling is strictly empirical and
intended only to capture dominant effects over a limited range of variables, or
it is based on arguments from more or less elementary dislocation dynamics
and then calibrated by choosing unknown parameters so as to fit selected data.
In principle, the more detailed the physical modeling, the more accurate the
constitutive equations will be and the more extensive their region of applica-
bility. In either case, the usual practice today involves a considerable amount
of empiricism, and at least to some extent, it is always necessary to fit the analytical expressions to data, by choosing unknown constants for the best fit. It is
also common practice to express the flow rule in inverted form, rather than as
in (3.63). Thus, $s = \hat{s}(\kappa, T, \dot{\gamma}^p)$ rather than $\dot{\gamma}^p = \Gamma(s, T, \kappa)$.

The remainder of this chapter describes a characteristic selection of both
empirical and physical models that have been used in the last decade or so.
Where possible the models are given both in their original form and in modified
form with an evolving work hardening parameter, as has been set out in general
terms above.

4.2.1 Power Law Model

A simple form that combines both work hardening and rate hardening in a single
function has been used by Molinari and Clifton (1987), Gioia and Ortiz (1996),
and others.

$$s = \pm \tau_0 |\gamma|^n T^{-v} |\dot{\gamma}|^m, \tag{4.18}$$

where the exponents n, v, and m are all positive numbers, and s and $\dot{\gamma}$ have
the same sign. As written, the flow stress exhibits no definite yield, but if the
work hardening exponent is small, the stress–strain curve has a sharp knee
that can be taken to simulate yield. The formula has actually been written for
unidirectional straining, and if stress reversals are to be introduced, plastic strain
must be understood to be the equivalent plastic strain. Similarly, as written, the
stress vanishes with strain rate, which is a completely unphysical value for a
solid, but for strong alloys it is often observed experimentally that a plot of
the logarithm of stress versus the logarithm of strain rate at a fixed strain and
temperature is linear over several decades of strain rate. Thus, experimental
data often show a power law correlation between stress and strain rate over a
significant data set. As actually used in problems, neither strain nor strain rate

is close to zero. Consequently the three principal effects of work hardening (flow stress increases with strain), thermal softening (flow stress decreases with increasing temperature), and strain rate hardening (flow stress increases with strain rate) are all captured by (4.18) in an elementary way.

Furthermore, Equation (4.18) suggests the generalization

$$s = \pm\kappa g(T)h(|\dot\gamma^P|). \tag{4.19}$$

In Equation (4.19) the work hardening parameter, κ, should evolve according to some constitutive function $\xi(s, T, \kappa)$ as described previously, $g(T)$ is a thermal softening function, normalized so that $g(T_0) = 1$, and $h(|\dot\gamma^P|)$ is a rate hardening term, normalized so that $h(0) = 1$. According to the previous description, the function, h, and the rate function, Γ, must be related according to $\Gamma(s, T, \kappa) = \pm h^{-1}[|s|/\kappa g(T)]$, where h^{-1} denotes the function inverse to h. Once again a constitutive function has been selected to be self-similar in its arguments. Of course, $h^{-1}(1) = 0$ and increases with increasing argument, so the rate effect follows from overstress, or an excess of flow stress over the quasi-static yield stress. Equation (4.19) is entirely empirical, but it allows considerable flexibility in fitting data over an extended range.

4.2.2 Litonski's Model

Litonski (1977) suggested a flow law of the (4.19) type with the particular form

$$s = C(\gamma_0 + \gamma^P)^n[1 - a(T - T_0)](1 + b\dot\gamma^P)^m. \tag{4.20}$$

As usual, the response function must be calibrated to represent a particular quasi-static work hardening curve at a particular reference temperature. The linear thermal softening may seem overly specialized, but later on it will be seen that the initial rate of softening often dominates the response in the early stages before a shear band is fully formed. Thus, the coefficient $a \equiv -g_T(T_0)$, which determines the initial rate, has great importance, but the remaining shape of the softening function, $g(T)$, has less importance. Later in the process, after the band is fully formed, the shape of the softening function has much greater effect, but then the conditions in the band become so extreme that no constitutive data are available for calibration anyway.

At the high rates of deformation that are characteristic of shear bands, the product $b\dot\gamma^P$ may be expected to be much greater than one, so the leading 1 in the last factor of (4.20) can often be safely ignored. The power law form, where $m < 0.1$ for most metals, is suggested by log–log plots of stress versus strain rate at constant strain, as mentioned previously. In terms of an evolving work hardening parameter, Equation (4.20) may be rewritten as

$$s = \kappa(1 - a\vartheta)(b\dot{\gamma}^{p})^{m} \tag{4.21}$$

where $\vartheta = T - T_0$ and the leading 1 in the last factor has been left out for the reason stated above.

4.2.3 Johnson–Cook Model

Johnson and Cook (1983) proposed the following empirical formula to represent dynamic plastic flow:

$$s = (A + B\gamma^{n})(1 + m \ln \dot{\gamma})[1 - (T^{*})^{v}], \tag{4.22}$$

where A, B, m, n, and v are constants. The term T^{*}, given by

$$T^{*} = (T - T_0)/(T_m - T_0), \tag{4.23}$$

is just a linear interpolating factor that vanishes at a reference temperature, T_0, and has the value 1 at the melting (or other designated) temperature of the material, T_m. (In the original work the symbols C, m were used rather than m, v, respectively, as presented here.)

Johnson and Cook also published tables of parameters for many common materials, and because of easy access to these data, Equation (4.22) is one of the most widely used flow laws for large-scale computations today. The leading factor represents work hardening and may be replaced by the work hardening parameter, κ, as shown in the first section of this chapter. The second factor is essentially the same as the power law for small rate hardening exponents, actually for $(m \ln \dot{\gamma}^{p})^{2} \ll 1$, because the strain rate term in (4.21) becomes $(\dot{\gamma}^{p})^{m} = e^{m \ln \dot{\gamma}^{p}} = 1 + m \ln \dot{\gamma}^{p} + O(m \ln \dot{\gamma}^{p})^{2}$ when expanded as a power series.

Except for the value $v = 1$, the thermal term in (4.22) must be regarded as unsatisfactory because the rate of thermal softening either vanishes at the reference temperature (if $v > 1$) or is infinite there (if $v < 1$). Either case must be regarded as unphysical because the reference temperature is unlikely to be exceptional in any way. Even if the thermal factor is linear (the most commonly used case), the initial rate of thermal softening is determined solely by the melting temperature, which again is physically unlikely, but to obtain a better initial rate of softening, an arbitrary choice for T_m may be used. An example is shown in Figure 4.4, where the flow stress in uranium decreases linearly until a phase change is reached. Because the initial rate of thermal softening is one of the most critical parameters in localization problems, it is important that the model should have sufficient flexibility to mimic the correct initial behavior.

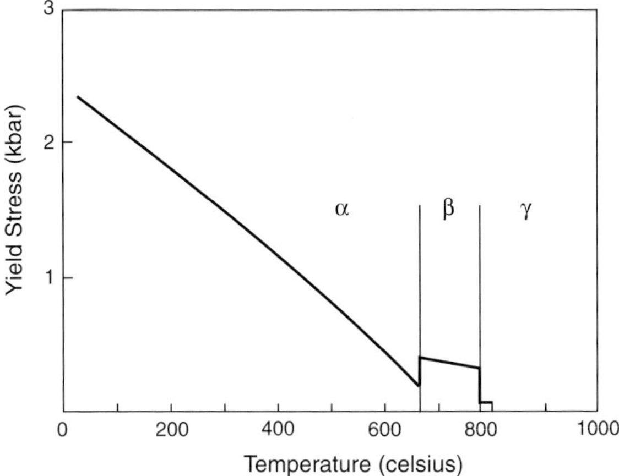

Figure 4.4. Yield stress for uranium as a function of temperature. (Adapted from Holden 1958, with permission from Addison-Wesley–Longman.)

Although widely used, Equation (4.22) is much the same as the Litonski law in the form of either (4.20) or (4.21), and all may be regarded as special cases of (4.19).

4.2.4 Zerilli–Armstrong Models

Zerilli and Armstrong (1987) proposed flow laws for both fcc and bcc polycrystalline materials. Although their paper is phrased in the language of dislocation mechanics, there is still a significant empirical component. Reduced to its essentials, their argument is as follows. Because plastic slip is thought to be a thermally activated process, strain rate should satisfy an Arrhenius law

$$\dot{\gamma} = \Gamma_0 e^{-G/kT}, \tag{4.24}$$

where Γ_0 is a constant, G is an activation energy that depends on stress and temperature, k is Boltzmann's constant, and T is absolute temperature. The flow stress is assumed to have both athermal and thermal components,

$$s = s_a + s_{\text{th}}, \tag{4.25}$$

where the thermal component of stress may be expressed as

$$s_{\text{th}} = s_{\text{th}_0} e^{-\beta T} \tag{4.26}$$

and the exponential factor, β, depends on strain rate

$$\beta = \beta_0 - \beta_1 \ln(\dot{\gamma}^p / \Gamma_0). \tag{4.27}$$

In (4.27), β_0, β_1, and Γ_0 are constants to be determined experimentally, but Γ_0 is usually taken to be one. To be self-consistent, the activation energy must be given by

$$G = \frac{kT(\beta - \beta_0)}{\beta_1} = \frac{k}{\beta_1} \ln \frac{s_{th_0}}{s_{th}} - \frac{kT\beta_0}{\beta_1}. \tag{4.28}$$

They then argue that for bcc metals, work hardening and thermal activation are independent processes, but in fcc metals the processes are strongly coupled. The net result is that for the two cases the flow stresses are expressed as

$$s = s_0 + C(\gamma^P)^{1/2}e^{-\beta T} \qquad \text{for fcc,}$$
$$s = s_0 + C(\gamma^P)^n + Be^{-\beta T} \quad \text{for bcc.} \tag{4.29}$$

In (4.29) the expression for β is the same as in (4.27), and s_0, B, C, and n are constants. In addition, the constant s_0 may be affected by the average grain size of the polycrystalline material. In actual application, although the forms have some theoretical basis, equations (4.29) are simply used empirically to fit data.

Expressed in terms of a single work hardening parameter, it is natural to write

$$s = s_0 + \kappa e^{-\beta T} \quad \text{for fcc,}$$
$$s = \kappa + Be^{-\beta T} \quad \text{for bcc.} \tag{4.30}$$

The strain hardening indicated by the power law in (4.29) may be mimicked by assuming an appropriate evolutionary law for κ, such as

$$\dot{\kappa} = \xi(s, T, \kappa) = n \left(\frac{\kappa}{C}\right)^{-1/n} \kappa \Gamma(s, T, \kappa), \tag{4.31}$$

where C is the same constant as in (4.29). In arriving at (4.31) for both fcc and bcc materials, the strain has been eliminated from the slope of the stress–strain curves in (4.29), as was done in the last section, and the function, Γ, for the plastic strain rate is found from the inverse of (4.30) with the aid of (4.27).

$$\Gamma(s, T, \kappa) = \begin{cases} \Gamma_0 e^{\beta_0/\beta_1} \left(\dfrac{s - s_0}{\kappa}\right)^{1/\beta_1 T} ; & s_0 < s < s_0 + \kappa, \text{ fcc,} \\[6mm] \Gamma_0 e^{\beta_0/\beta_1} \left(\dfrac{s - \kappa}{B}\right)^{1/\beta_1 T} ; & \kappa < s < \kappa + B, \text{ bcc.} \end{cases} \tag{4.32}$$

Because Equation (4.31) has the form $\xi = \phi(\kappa)\Gamma$, it cannot reproduce history effects, as noted previously, but modifications similar to (4.10) will do so. In any particular case, it would be necessary to decide what effect is to be modeled and to modify (4.31) appropriately.

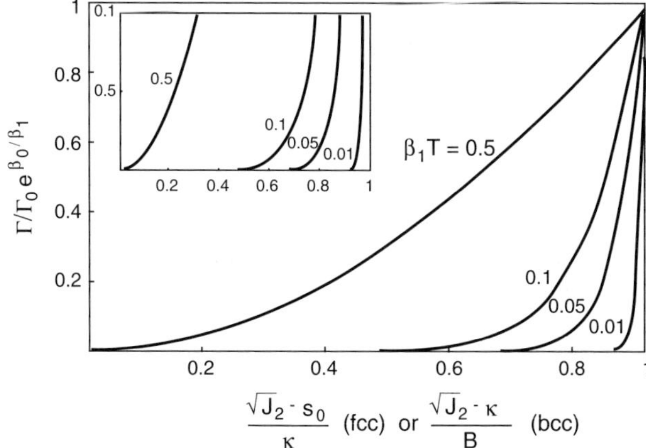

Figure 4.5. Plastic strain rate according to the Zerilli–Armstrong flow law of Equation (4.32). At low temperatures, the strain rate appears to increase weakly as the stress invariant exceeds a lower limit, increasing rapidly as it approaches an upper limit. In fact the calibrating strain rate, $\Gamma_0 \exp(\beta_0/\beta_1)$, is usually so large that only the bottom of the curve is ever used. The inset shows a blowup of the curve for smaller strain rates.

Equation (4.32) has the appearance of an overstress model for plastic strain rate with the plastic flow vanishing at $s = s_0$ and $s = \kappa$ in the two cases. Published values for the coefficient β_1 (in units of reciprocal degrees Kelvin) seem to be approximately in the range $10^{-4} < \beta_1 < 5 \times 10^{-4}$, with β_0 some 15 to 25 times larger (Zerilli and Armstrong 1987, 1990), so that $\beta_1 T$, which is effectively the strain rate sensitivity, is small, usually much less than one, but $\exp(\beta_0/\beta_1)$ may be $\mathcal{O}(10^6)$ to $\mathcal{O}(10^{11})$. Thus, the final exponent in (4.32) is large so the right-hand side in each case is small until there is a knee in the curve with a rapid upturn shortly before the quantity in brackets reaches one. The behavior of (4.32) is illustrated in Figure 4.5. Even with $\Gamma_0 = 1$, the preexponential factor is also large, however, so the useful range of the function may lie closer to the horizontal axis where the shape of the function is shown in the inset.

Although (4.32) no longer has the appearance of an Arrhenius type of dependence, in fact the exponentiated term is very close to an exponential of the expected type. A good approximation to (4.32) may be found after first noting that for $|x| < 1$ and $|hx^2| \ll 1$, the expression $y = (1 + x)^h$ has the asymptotic representation $y \sim e^{hx}[1 + \mathcal{O}(hx^2)]$. [Take logarithms of both sides of the expression to get $\ln y = h\ln(1 + x) = hx + h(-1/2x^2 + \cdots)$. Then, the result follows after reexponentiation because $\exp h(-1/2x^2 + \cdots) = 1 + \mathcal{O}(hx^2)$.]

When applied to (4.32), this approximation leads to

$$\Gamma\,(s,\,T,\,\kappa)$$

$$= \Gamma_0 \begin{cases} \exp\left[\dfrac{s - s_0 - \kappa(1 - \beta_0 T)}{\kappa\beta_1 T}\right]; & \left|1 - \dfrac{s - s_0}{\kappa}\right| \ll \sqrt{\beta_1 T},\ \text{fcc,} \\[3ex] \exp\left[\dfrac{s - \kappa - B(1 - \beta_0 T)}{B\beta_1 T}\right]; & \left|1 - \dfrac{s - \kappa}{B}\right| \ll \sqrt{\beta_1 T},\ \text{bcc.} \end{cases}$$

$$(4.33)$$

Even for the minimum values in the two cases, $s = s_0$ or $s = \kappa$, the approximations are still very close because Γ will be exponentially small instead of vanishing exactly, as it should.

4.2.5 Bodner–Partom Model

Over a period of thirty years or so S. R. Bodner, in collaboration with various colleagues, has developed a unified theory of viscoplastic flow that does not use an explicit yield surface at all (e.g., see Bodner 1987). Like the Zerilli–Armstrong models, however, strain rate makes a rapid transition, in this case from being exponentially small to rapid growth. Plastic flow is assumed to follow (3.59), written in the form $\boldsymbol{D}^p = 1/2\Gamma(J_2, T, \kappa)(\boldsymbol{S}/\sqrt{J_2})$, or as written in the original papers (with some inconsequential changes of notation):

$$\Gamma = 2\Gamma_0 \exp[-1/2(\kappa^2/J_2)^n]. \qquad (4.34)$$

In (4.34), Γ_0 is a constant, $J_2 = -\mathrm{II}_S = 1/2\,\mathrm{tr}\,\boldsymbol{S}^2$ (as defined previously in Chapter 3), and n depends on temperature, $n = (a/T) + b$. The work hardening parameter, κ, which here is referred back to a flow stress in shear rather than tension as in the original work, evolves according to the equation

$$\dot{\kappa} = \alpha(\kappa_1 - \kappa)(\boldsymbol{S}:\boldsymbol{D}^p). \qquad (4.35)$$

According to (4.35), with $\kappa_1 > \kappa_0$, the work hardening parameter eventually saturates with plastic work and may be expressed as $\kappa = \kappa_1 - (\kappa_1 - \kappa_0)\exp(-\alpha W^p)$.

The behavior of (4.34) is illustrated in Figure 4.6. For a given value of n, note that the plastic strain rate switches from being exponentially small to rapid growth. However, the maximum plastic strain rate is $\Gamma_{\max} = 2\Gamma_0$, and so Γ_0 may have to be chosen to have different values for low and high strain rate applications. The effective strain rate sensitivity may be calculated to be $\frac{\delta s/s}{\delta\Gamma/\Gamma} = \frac{(s/\kappa)^{2n}}{2n}$, so like the Zerilli–Armstrong model, it depends on temperature, but here it also depends on the stress state.

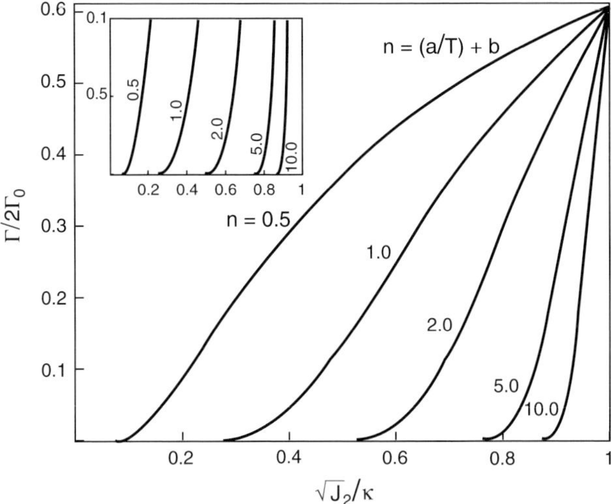

Figure 4.6. Plastic strain rate according to the Bodner–Partom flow law of Equation (4.34). At low temperatures (large n), the strain rate increases weakly at first as the stress invariant increases with a rapid upturn later. Temperature has a powerful effect on the strain rate. The inset shows a blowup of the curve for smaller strain rates.

4.2.6 MTS Model

MTS stands for mechanical threshold stress and refers to an internal variable (or variables) that evolves as plastic deformation progresses. Models of this kind were described by Kocks, Argon, and Ashby (1975) and have been developed and applied by a number of researchers since. The basic idea, which is essentially the same as that used by Zerilli and Armstrong, is that the stress required to move a single dislocation past an obstacle is composed of both thermal and athermal components. Moreover, the strain rate for the thermal component obeys an Arrhenius law. Unlike the model of Zerilli and Armstrong, however, the MTS model tries to avoid using plastic strain as an internal variable. The flow stress (not the strain rate, which would have been the natural choice in this book) is written as a sum of terms (e.g., see Follansbee 1989):

$$(|s| - s_a)^r = \sum_{i=1}^{n} [s_i(|\dot{\gamma}^P|, T)\hat{s}_i]^r. \tag{4.36}$$

There are as many terms in the sum as there are conceived to be different types of impediments to the motion of dislocations, and the power, r, is chosen empirically, $1 \leq r \leq 2$. The term s_a, called an athermal stress, may be either a constant or an evolutionary internal variable, and the $\hat{s}_i$ are called structural variables that also evolve as deformation progresses. The remaining functions of strain rate and temperature, $s_i(|\dot{\gamma}^p|, T)$, which take on values between zero and one, are obtained from the Arrhenius law (4.24) with activation energies of the form

$$G = \frac{k}{\beta}\left[1 - \left(\frac{|s| - B}{\hat{s}}\right)^p\right]^q, \quad 0 < p \le 1, \, 1 \le q \le 2, \tag{4.37}$$

where the constants p and q are purely empirical. Equations (4.36), (4.37), and (4.24), written in terms of a single internal variable, now designated κ rather than $\hat{s}$, and then combined, give a form that may be compared with Equation (4.30), namely

$$|s| = B + \kappa\left[1 - \left(\beta T \ln \frac{\Gamma_0}{\dot{\gamma}^p}\right)^{1/q}\right]^{1/p}. \tag{4.38}$$

As applied by Follansbee and Kocks (1988), the evolution of κ is to be found from an expression of the form

$$\vartheta = \frac{d\kappa}{d\gamma^p} = \mathcal{F}(\kappa, T, \dot{\gamma}^p). \tag{4.39}$$

The function $\mathcal{F}$ has been chosen in various forms, such as one that decreases linearly with κ and vanishes when κ reaches a saturation value. Put in evolutionary form, (4.39) implies

$$\dot{\kappa} = \mathcal{F}[\kappa, T, \Gamma(s, T, \kappa)]\Gamma(s, T, \kappa) = \xi(s, T, \kappa), \tag{4.40}$$

where $\Gamma(s, T, \kappa)$ is to be found by inversion of Equation (4.38).

A recent paper by Nemat-Nasser and Li (1998) develops a more elaborate version of (4.38) to model the dynamic response of Cu. In their version, B, κ, and Γ_0 each depend on the equivalent plastic strain, and B and κ depend on T, as well.

Common choices for the exponents p, q are $p = 2/3$ and $q = 2$. The inverse of (4.38) in generic form may be written

$$\dot{\gamma} = \Gamma_0 \exp\left\{-\frac{1}{\beta T}\left[1 - \left(\frac{|s| - B}{\kappa}\right)^{2/3}\right]^2\right\}. \tag{4.41}$$

Equation (4.41) is plotted in Figure 4.7 for various values of βT. Because $\beta = \mathcal{O}(10^{-5})$, according to data given by Follansbee and Kocks (1988), the strain rate effectively vanishes when $|s| = B$ but increases rapidly for $|s| > B$. Like the Bodner–Partom model, at large overstress the strain rate saturates at a prescribed maximum of Γ_0, but at a value that is something like two orders of magnitude larger than the maximum that is customarily used in the Bodner–Partom model, that is, $\mathcal{O}(10^7)$ instead of $\mathcal{O}(10^5)\,\mathrm{s}^{-1}$. As a consequence, for most applications the exact value of Γ_0 is of little consequence. The effective

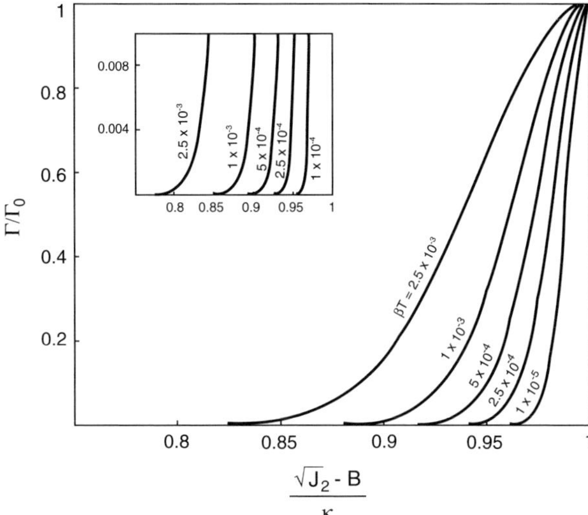

Figure 4.7. Plastic strain rate according to the MTS flow law of Equation (4.41). As with the Zerilli–Armstrong law, the calibrating strain rate is usually so large that only the bottom of the curve is actually used. The inset shows a blowup of the curve for smaller strain rates.

strain rate sensitivity is again proportional to βT, but it is modified by a complex function of $(s - B)/\kappa$.

4.2.7 Anand's Model

Anand (1985), followed later by Anand and Brown (1987) and Brown, Kim, and Anand (1989), developed constitutive laws specifically for application to hot working of metals. The key observation in this case seems to be that although there is no definite yield surface, a steady flow stress, τ, is reached at constant strain rate and temperature, and the use of the Zener–Hollomon parameter, $Z \equiv \dot{\gamma}^p e^{G/kT}$ (Zener and Hollomon 1944) provides a good, one-parameter correlation of the type $\tau = f(Z)$. In this expression G is a constant activation energy, and k is Boltzmann's constant. Furthermore, they state that the empirical relation $\tau = \tau_0 \sinh^{-1}[(Z/\Gamma_0)^m]$, where τ_0, Γ_0, and m are constants, is known to provide good correlation of data at high temperatures and moderate strain rates. To extend the steady-state relationship to unsteady conditions and to express the relationships in a form for applications to isotropic materials in multidimensions, they assumed (in the notation of this book) that

$$\Gamma(\sqrt{J_2}, T, \kappa) = \Gamma_0 e^{-G/kT} [\sinh(\sqrt{J_2}/\kappa)]^{1/m}, \qquad (4.42)$$

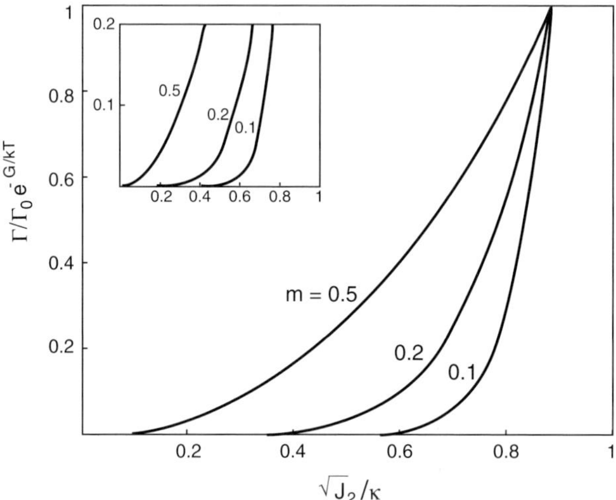

Figure 4.8. Plastic strain rate according to Anand's flow law of Equation (4.42). Although similar in appearance to the Zerilli–Armstrong curves, here the curves are parameterized by the constant m rather than a function of temperature. The inset shows a blowup of the curve for smaller strain rates.

where the evolutionary law for κ is given by

$$\dot{\kappa} = \xi(\sqrt{J_2}, T, \kappa) = h_0 \left| 1 - \frac{\kappa}{\hat{\kappa}} \right|^{a-1} \left(1 - \frac{\kappa}{\hat{\kappa}} \right) \Gamma(\sqrt{J_2}, T, \kappa), \quad a \geq 1,$$

$$\hat{\kappa} = \kappa_0 \left[\left(\frac{\Gamma}{\Gamma_0} \right) e^{G/kT} \right].$$

$$(4.43)$$

In (4.43), h_0 and κ_0 are constants. For a fixed temperature and strain rate, note that $\hat{\kappa}$ takes on a fixed value. Therefore, κ saturates at the fixed value $\hat{\kappa}$ where $\dot{\kappa} = 0$, and $\sqrt{J_2}$ takes on a saturation value as determined by (4.42). Also note that in this model κ itself may be either increasing or decreasing because $\hat{\kappa}$ may change with changing temperature and strain rate, thus changing the sign of the right-hand side of (4.43). Figure 4.8 shows how the strain rate varies with the imposed stress.

4.3 Concluding Remarks

With the introduction of only one internal variable, there are four functions that may be constructed so as to match given data in simple shearing over a wide range of temperatures and strain rates: the yield function $\hat{f}(s, T)$, the rate of work hardening $\xi(s, T, \kappa)$, the induced rate of plastic strain $\Gamma(s, T, \kappa)$, and the generalized force $Q(s, T, \kappa)$. Even though these functions cannot be completely arbitrary because they must be mutually consistent with each other and with the requirements of thermodynamics, it would seem that there is still ample scope to fit a wide variety of data, including simple modeling of recovery. However, it must be noted that because hardening mechanisms in metals are numerous and complex, the use of only one internal variable is probably appropriate only for the simplest of materials such as high purity, polycrystalline fcc metals. To use only one internal variable in more complex cases, it must be argued that under the range of conditions to be modeled, one mechanism dominates, and all others are essentially frozen. In fact, the full range of capabilities of the theory sketched out here has never been tested.

There are broad similarities in the four flow models represented in Figures 4.5–4.8. The differences may become pronounced when the models are calibrated for particular materials. Usually the actual range of application is only a fraction of that shown in the figures, and the actual physical range may be greatly expanded or contracted by the details of the calibration to data. Because each flow law depends on at least one structural parameter (κ, representing work hardening in this chapter), the actual strain rate response will depend strongly on the evolution of the structure.

Models 3–7 have undergone further development and are active today. The models described in this chapter are representative only, and many others may be found in the literature (e.g., Steinberg and Lund 1989, Bammann 1990, or Miller 1987). Recent references may be found in the current literature.

5

One-dimensional problems, part I: General considerations

Adiabatic shear bands are extremely thin in one direction compared with their extent in the other two perpendicular directions. Variations of temperature and velocity in the thin direction tend to be extremely rapid when compared with variations within the plane of the band itself. As a consequence of this morphology, it is possible to understand a great deal about the dynamics of band formation simply from examination of the one-dimensional equations, repeated here in the simplified form as developed in Section 3.3 of Chapter 3.

$$
\begin{aligned}
\rho\, v_t &= s_y, & &\text{momentum;} \\
\rho c T_t &= (k T_y)_y + \beta s \dot{\gamma}^p, & &\text{energy;} \\
\dot{s} &= \mu(v_y - \dot{\gamma}^p), & &\text{elasticity;} \\
\dot{\gamma}^p &= f(s, \kappa, T), & &\text{flow law;} \\
\dot{\kappa} &= \xi(s, \kappa, T), & &\text{work hardening.}
\end{aligned}
\tag{5.1}
$$

Qualitative and quantitative solutions of these equations, or more often a specialization of the equations, will be developed in this and the following two chapters so as to bring out some of the most important analytical features of shear bands.

5.1 Homogeneous Solutions and the Reduction to Rigid-Plastic Material

Homogeneous shearing is the name given to solutions of (5.1) when strain rate is constant, and stress, temperature, and work hardening depend only on time but not on spatial variables. Because of adiabatic heating, that is, heating that arises from the plastic work, and because of the general tendency for materials to soften at higher temperatures, homogeneous solutions generally

98

have a maximum in stress followed by a region in which stress decreases slowly with increasing strain. It is important to have an understanding of these solutions as background and predecessor to instability and localization because solutions generally follow the homogeneous solution closely at first and then branch off rapidly to exhibit localized behavior.

Because the elastic shear modulus is much larger than the elastic yield stress, often by nearly two orders of magnitude, it is possible without significant error to ignore the elastic deformation and to treat the material as if it were rigid plastic, as pointed out by Wright and Walter (1989). To see this, consider steady, homogeneous shearing of a material that initially is at the quasi-static elastic limit of the material, s_0, the reference value of work hardening, κ_0, and the reference temperature, T_0. Thus, the velocity profile is given by $v = \dot{\gamma}_0 y$ (where $\dot{\gamma}_0$ is a constant, applied strain rate), and there is no y dependence in the other variables, s, κ, and T. Let $\beta = 1$, and nondimensionalize the equations by letting

$$\tau = \dot{\gamma}_0 t, \quad \theta = \rho c(T - T_0)/s_0, \quad \sigma = s/s_0, \quad \lambda = \kappa/\kappa_0 \qquad (5.2)$$

so that in nondimensional form they reduce to

$$\theta_\tau = \sigma \hat{f}(\sigma, \lambda, \theta),$$

$$\varepsilon \sigma_\tau = 1 - \hat{f}(\sigma, \lambda, \theta), \qquad (5.3)$$

$$\lambda_\tau = \hat{\xi}(\sigma, \lambda, \theta),$$

where $\varepsilon = s_0/\mu$ is a small parameter, $\varepsilon \ll 1$, and $\hat{f}(\sigma, \lambda, \theta) = \dot{\gamma}_0^{-1} f(\sigma s_0, \lambda \kappa_0, T_0 + s_0\theta/\rho c)$ and $\hat{\xi}(\sigma, \lambda, \theta) = (\kappa_0 \dot{\gamma}_0)^{-1}\xi(\sigma s_0, \lambda \kappa_0, T_0 + s_0\theta/\rho c)$. At the onset of yield, $\sigma(0) = \lambda(0) = 1$, $\theta(0) = 0$, and there is no plastic strain rate, so $\hat{f}(1, 1, 0) = 0$. If $\beta \neq 1$, but still constant, it can be absorbed into the scaling for temperature, $\theta = \rho c(T - T_0)/\beta s_0$, so that equations (5.3) remain unchanged.

5.1.1 Initial Boundary Layer: General Description

Equations (5.3) show a classic boundary layer at the initial time (e.g., see Bender and Orszag 1978). A regular perturbation in the small parameter ε (i.e., $\sigma = \sigma^0 + \varepsilon\sigma^1 + \cdots$) at lowest order leads to the equations

$$\theta_\tau^0 = \sigma^0 \hat{f}(\sigma^0, \lambda^0, \theta^0),$$

$$0 = 1 - \hat{f}(\sigma^0, \lambda^0, \theta^0), \qquad (5.4)$$

$$\lambda_\tau^0 = \hat{\xi}(\sigma^0, \lambda^0, \theta^0).$$

The solution of the second equation of (5.4) for stress as a function of the other two variables corresponds to a rigid-plastic material (because $\varepsilon \to 0$ as $\mu \to \infty$) and may be expressed as $\sigma^0 = \hat{s}(\lambda^0, \theta^0)$. Substitution of this result into the other two equations results, at lowest order, in a pair of first-order ODEs:

$$\frac{d\theta_0^{\text{out}}}{d\tau} = \theta_\tau^0 = \hat{s}(\lambda^0, \theta^0)\hat{f}(\sigma^0, \lambda^0, \theta^0) = \hat{s}(\lambda^0, \theta^0),$$

$$\frac{d\lambda_0^{\text{out}}}{d\tau} = \lambda_\tau^0 = \hat{\xi}[\hat{s}(\lambda^0, \theta^0), \lambda^0, \theta^0] = \hat{\hat{\xi}}(\lambda^0, \theta^0).$$

$$(5.5)$$

The system is now one order lower than the original system, or the solution of (5.5) cannot satisfy all three of the original initial conditions, and therefore it must yield only the outer asymptotic solution to (5.3), as anticipated by the new notation. As such its initial values must correspond to the outer limit of the boundary layer or inner solution, which has yet to be determined.

The boundary layer equations may be found by making a change of the independent variable, $\eta = \tau/\varepsilon$, so that Equations (5.3) become

$$\theta_\eta = \varepsilon\sigma\hat{f}(\sigma, \lambda, \theta),$$

$$\sigma_\eta = 1 - \hat{f}(\sigma, \lambda, \theta),$$

$$\lambda_\eta = \varepsilon\hat{\xi}(\sigma, \lambda, \theta).$$

$$(5.6)$$

At lowest order, the first and third equations have constant solutions, $\theta_0^{\text{in}} = 0$ and $\lambda_0^{\text{in}} = 1$, and the second equation becomes

$$\frac{d\sigma_0^{\text{in}}}{d\eta} = 1 - \hat{f}\left(\sigma_0^{\text{in}}, 1, 0\right).$$

$$(5.7)$$

At the initial value for stress, $\sigma_0^{\text{in}}(0) = 1$, the function $\hat{f}(1, 1, 0) = 0$, so $d\sigma_0^{\text{in}}(0)/d\eta = 1$, and the stress begins to increase rapidly (essentially elastically) toward the value $\bar{\sigma}$ at which $\hat{f}(\bar{\sigma}, 1, 0) = 1$ where the right-hand side of (5.7) vanishes. The asymptotic values of λ_0^{in}, θ_0^{in} at large η are still $(1, 0)$ to lowest order so the initial values for problem (5.5) are $\lambda_0^{\text{out}}(0) = 1$ and $\theta_0^{\text{out}}(0) = 0$. These values are consistent with the inner limit of the outer solution for stress, $\hat{s}(1, 0) = \bar{\sigma}$. The net result is that after a rapid transient, the stress, work hardening, and temperature all behave as if the material deformed adiabatically as a rigid-plastic material.

5.1.2 Initial Boundary Layer: A Simple Example

A simple example for a nonwork hardening material may be used to illustrate the transient elastic behavior in homogeneous, adiabatic shearing. Suppose that plastic flow is described by $s = \kappa_0[1 - a(T - T_0)](1 + b\dot{\gamma}^p)^m$, where κ_0 is the initial flow stress, a and b are constants, and strain rate sensitivity, m, is a small number. With a small parameter defined by $\varepsilon = \kappa_0/\mu$, nondimensional variables defined by $\tau = \dot{\gamma}_0 t$, $\sigma = s/\kappa_0$, $\theta = \rho c(T - T_0)/\kappa_0$, $\dot{\gamma} = \dot{\gamma}^p/\dot{\gamma}_0$ and nondimensional constants by $\alpha = \kappa_0 a/\rho c$ and $\beta = b\dot{\gamma}_0$, the equations corresponding to (5.3) are

$$\varepsilon\sigma_\tau = 1 - \dot{\gamma}, \quad \sigma(0) = 1,$$
$$\theta_\tau = \sigma\dot{\gamma}, \quad\quad \theta(0) = 0,$$

$$(5.8)$$

where

$$\dot{\gamma} = \frac{1}{\beta}\left(\frac{\sigma}{1 - \alpha\theta}\right)^{1/m} - \frac{1}{\beta}.$$

For the zeroth-order outer solution, according to the first equation of (5.8), set $\dot{\gamma} = 1$, to get

$$\sigma_0^{\text{out}} = (1 + \beta)^m\left(1 - \alpha\theta_0^{\text{out}}\right),$$

$$\frac{\partial\theta_0^{\text{out}}}{\partial\tau} = (1 + \beta)^m\left(1 - \alpha\theta_0^{\text{out}}\right),$$

$$(5.9)$$

with the solution

$$1 - \alpha\theta_0^{\text{out}} = \left[1 - \alpha\theta_0^{\text{out}}(0)\right]e^{-\alpha(1+\beta)^m\tau}.$$

$$(5.10)$$

The initial value $\theta_0^{\text{out}}(0)$ must come from the outer limit of the inner solution and may be anticipated to be zero from the general case.

With the change of variable $\tau = \varepsilon\eta$ in (5.8), the equations for the inner solution become

$$\frac{d\sigma_0^{\text{in}}}{d\eta} = 1 - \frac{1}{\beta}\left[\left(\frac{\sigma_0^{\text{in}}}{1 - \alpha\theta_0^{\text{in}}}\right)^{1/m} - 1\right] = \frac{1 + \beta}{\beta} - \frac{\left(\sigma_0^{\text{in}}\right)^{1/m}}{\beta}; \quad \sigma_0^{\text{in}}(0) = 1,$$

$$\frac{d\theta_0^{\text{in}}}{d\eta} = 0; \quad\quad\quad\quad\quad\quad\quad\quad\quad\quad\quad\quad\quad\quad\quad\quad\quad \theta_0^{\text{in}}(0) = 0,$$

$$(5.11)$$

where the solution for the second equation of (5.11), $\theta_0^{\text{in}} = 0$, has been substituted into the first equation of (5.11). A further consequence is $\theta_0^{\text{out}}(0) = \theta_0^{\text{in}}(\infty) = 0$. If we now let $y = \sigma_0^{\text{in}}/(1 + \beta)^m$, (5.11) becomes

$$\frac{dy}{1 - y^{1/m}} = \frac{(1 + \beta)^{1-m}}{\beta}\,d\eta; \quad y(0) = (1 + \beta)^{-m} \leq y \leq 1. \quad (5.12)$$

Equation (5.12) must be integrated numerically for best accuracy, but because m is a small number, an approximate solution may be found by letting

$y = 1 - z; 1 - (1 + \beta)^{-m} = z(0) \geq z \geq 0$. Now note that if z is small (actually $|z| \ll \sqrt{2m}$), then $y^{1/m} = (1 - z)^{1/m} = e^{-z/m}[1 - \mathcal{O}(z^2/2m)]$ so that (5.12) may be approximated by

$$-\frac{e^{z/m} d(z/m)}{e^{z/m} - 1} = \frac{(1 + \beta)^{1-m}}{\beta} \frac{d\eta}{m},$$
(5.13)

which has the exact solution $e^{z/m} - 1 = [e^{z(0)/m} - 1] \exp\{-\beta^{-1}(1 + \beta)^{1-m} (\eta/m)\}$. After we once again use the approximation that connects y and z, the solution becomes

$$y = \frac{\sigma_0^{\text{in}}}{(1 + \beta)^m} \approx \left\{1 + \beta \exp\left[-\frac{(1 + \beta)^{1-m}}{\beta} \frac{\eta}{m}\right]\right\}^{-m}.$$
(5.14)

The complete solution is approximated by $\sigma \sim \sigma^{\text{in}} + \sigma^{\text{out}} - \sigma^{\text{int}}$ where the third term is called the intermediate solution and is given by $\sigma^{\text{int}} = \sigma^{\text{in}}(\infty) = \sigma^{\text{out}}(0)$ (e.g., see Bender and Orszag 1978). In dimensional terms at lowest order the stress and temperature are approximated by

$$\frac{s/\kappa_0}{(1 + b\dot{\gamma}_0)^m} \sim \left\{1 + b\dot{\gamma}_0 \exp\left[-\frac{(1 + b\dot{\gamma}_0)^{1-m}}{b\dot{\gamma}_0} \frac{\mu}{\kappa_0} \frac{\dot{\gamma}_0 t}{m}\right]\right\}^{-m}$$

$$+ \exp\left[-\frac{a\kappa_0}{\rho c}(1 + b\dot{\gamma}_0)^m \dot{\gamma}_0 t\right] - 1$$
(5.15)

$$a(T - T_0) \sim 1 - \exp\left[-\frac{a\kappa_0}{\rho c}(1 + b\dot{\gamma}_0)^m \dot{\gamma}_0 t\right].$$

In the first of (5.15) the last two terms cancel each other at $t = 0$, and the first and third cancel as $t \to \infty$ in the typical manner for inner–outer expansions. Note particularly the nondimensional grouping of terms, $a\kappa_0(1 + b\dot{\gamma}_0)^m/\rho c \equiv (\rho c)^{-1}(\partial s/\partial T)$. This combination appears over and over in the analysis of adiabatic shear bands, usually (but not always) evaluated at the reference temperature.

Figure 5.1 shows the comparison of the exact solution, as computed by the program Mathematica, and the asymptotic solution in (5.15) for the values $\alpha = 0.1$, $\beta = 10^7$, $\varepsilon = 0.01$, and $m = 0.02$, with $0 \leq \tau \leq 0.1$. The exact and asymptotic solutions are barely distinguishable and clearly bring out the rapid transition from mostly elastic to perfectly plastic behavior. The asymptotic solution, because of its explicit representation, also shows how the transitional behavior may be expected to change with the nondimensional parameters.

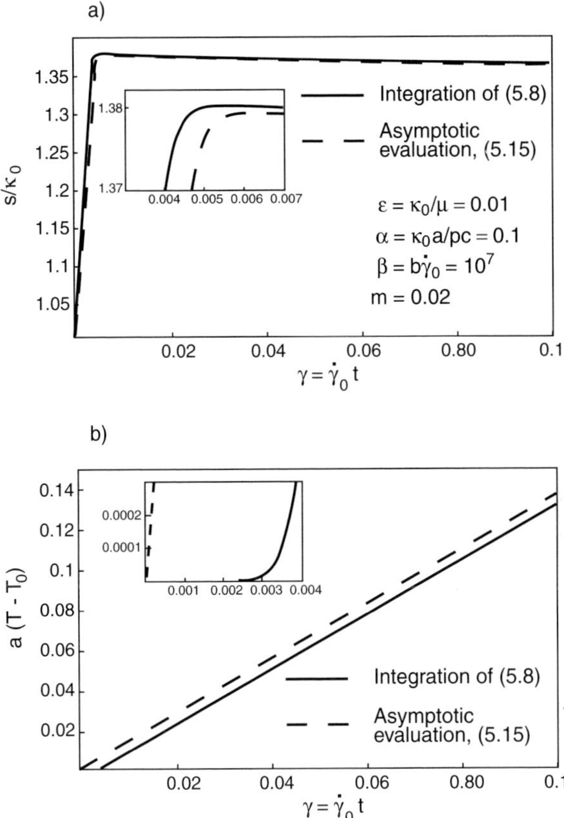

Figure 5.1. With monotonic shearing, the response quickly switches from essentially elastic to essentially plastic. Stress and temperature are shown in panels a and b, respectively, as computed from the ODEs and from an asymptotic evaluation. The inset in panel a shows a blowup of the response near the knee in the curve; that in panel b shows a blowup of the initial rise in temperature.

5.2 Steady Solutions

Equations (5.1) also admit steady solutions where all fields (velocity, stress, temperature, and work hardening) depend only on the spatial coordinate, but not on time. Solutions, as given by Wright (1987b) and by Bai and Dodd (1992), can be obtained as quadratures from reduced balance and constitutive laws. First note from the last equation of (5.1) that in a steady solution the work hardening must be saturated so that $\kappa_t = 0 = \xi(s, \kappa, T)$. After inversion of ξ, the work hardening parameter may be expressed as a function of stress and temperature, $\kappa = \hat{\kappa}(s, T)$. Furthermore, if the velocity is independent of time, balance of momentum demands that the stress must be independent of the y coordinate as well as time, and therefore it is constant, $s = \bar{s}$. Because stress is constant, the elastic strain rate vanishes, and according to the third equation of (5.1), the total strain rate is now equal to the plastic strain rate, which may be expressed as

$$v_y = \dot{\gamma}^p = f[\bar{s}, \hat{\kappa}(\bar{s}, T), T] = \hat{f}(\bar{s}, T). \tag{5.16}$$

Both velocity and work hardening depend only on stress and temperature, so all that remains is to find the distribution of temperature. If the thermal conductivity depends only on temperature, the solution follows easily by quadrature from the reduced energy equation:

$$\frac{d}{dy}\left(k\frac{dT}{dy}\right) + \bar{s}\hat{f}(\bar{s}, T) = 0. \tag{5.17}$$

A first integral may be found by multiplying through by kT_y and then integrating (5.17) on the coordinate, y. If integration extends from the center of the shear band, where the temperature is a maximum, to an arbitrary point, the first integral may be expressed as

$$\frac{1}{2}(kT_y)^2 = \bar{s}\int_T^{T_c} k(T')\hat{f}(\bar{s}, T')\,dT'. \tag{5.18}$$

Because the right-hand side of (5.18) depends only on temperature, a second integral may also be found to express the distance from the center of the band as a function of temperature:

$$y = \pm\frac{1}{\sqrt{2\bar{s}}}\int_T^{T_c}\frac{k(T')\,dT'}{\sqrt{\int_T^{'T_c} k(T'')\hat{f}(\bar{s}, T'')\,dT''}}. \tag{5.19}$$

The two signs refer to the two sides of the shear band and indicate that the solution is symmetric about the center of the band. The integrand is positive, and therefore the distance, y, and the temperature, T, have distinct monotonic relationships for each sign on the right-hand side.

The particle velocity relative to the center of the band may also be expressed by quadrature because the reduced energy equation, $(kT_y)_y + \bar{s}v_y = 0$, has an integral

$$v = -\frac{kT_y}{\bar{s}} = \pm\sqrt{\frac{2}{\bar{s}}}\sqrt{\int_T^{T_c} k(T')\hat{f}(\bar{s}, T')\,dT'}. \tag{5.20}$$

Again the two signs refer to the two sides of the band and indicate an antisymmetric distribution of velocity, which is monotonic with temperature on each side. Equations (5.19) and (5.20) together indicate that velocity increases monotonically with y all the way through the band from one side to the other.

Equations (5.18), (5.19), and (5.20) describe the structure of any steady solution for one-dimensional shearing relative to the center of the band where the maximum temperature occurs. Note that these solutions form a two-parameter family because they depend on assignment of the central temperature, T_c, and the stress, $\bar{s}$. Equivalently, a prescribed velocity or temperature could be assigned at a boundary, rather than one of the other two parameters. Whereas it is clear that for each choice of $\bar{s}$ and T_c there is a well-defined distribution of temperature and velocity, it is not so clear that there are well-defined solutions for other choices of the parameters. As an example, suppose that the temperature at the sides of a channel of width $2H$ is prescribed to be T_0. Thus, with the left side of (5.19) set equal to H, and the lower limit of the integral set equal to T_0, the equation represents a relationship between $\bar{s}$ and T_c. For each pair $(\bar{s}, T_c)$ that satisfies this relationship there is a well-defined boundary value of v that may be calculated from (5.20). Thus, there are pairs $(\bar{s}, \bar{v})$ that represent the stress and boundary velocity of a steady shearing solution in a channel of width, $2H$, and prescribed boundary temperature, T_0. However, it is not clear that an arbitrary choice of one allows the other to be found, or if it can be found, that the solution is stable. In fact, it usually seems to turn out that stress, $\bar{s}$, may be found as a unique function of velocity, $\bar{v}$, but that the relation is not monotonic. Thus, a given choice of stress may correspond to zero, one, or multiple values of velocity. Problems of this kind have been examined by Chen (1988), Chen, Douglas, and Malek-Madani (1989), Maddocks and Malek-Madani (1992), and Fleming, Olmstead, and Davis (2000).

Although the quadratures are exact and fully general, they are not particularly informative in their present form. Their greatest usefulness is not to solve a fixed boundary value problem, as described in the previous paragraph, but rather to describe a shear band as a boundary layer in the interior of a shearing motion. In this case, roughly speaking, the stress, $\bar{s}$, the exterior temperature, T_0, and driving velocity, $\bar{v}$, are all prescribed, but the width H is not given. It may seem odd to prescribe both the stress and velocity, in addition to the ambient temperature, but Equation (5.20) indicates that when this is done there is always a unique corresponding value of the central, maximum temperature because of the monotonic nature of the integral. Then (5.19) yields a unique spatial distribution for all field variables. In the next chapter, universal asymptotic results, based on Equations (5.19) and (5.20), will be given for a shear band considered as a boundary layer.

5.3 Change of Type, Regularization, and Embedded Change of Type

Material behavior that leads to extreme localization is often thought to have mathematical expression such that the equations of motion undergo a change of type. In a three dimensional, dynamic setting the usual hyperbolic signature of leading derivatives in a preferred coordinate system for a second-order partial differential equation (PDE) would be $(-1, +1, +1, +1)$, corresponding to one temporal and three spatial coordinates. When there is a change of type, the coefficients of the derivatives evolve in such a way that one of the spatial second derivatives changes signature from $+1$ to -1. In the general case the signature then becomes of mixed type $(-1, -1, +1, +1)$. Often the original problem is posed only in equilibrium, and then the signature in certain regions changes from $(+1, +1, +1)$ to $(-1, +1, +1)$ or from elliptic to hyperbolic. In the one dimensional, dynamical formulation of this book the change is from $(-1, +1)$ to $(-1, -1)$ or from hyperbolic to elliptic form in space and time. Then the initial value problem, which is appropriate for the hyperbolic form, becomes ill posed because it is "inappropriate" for the elliptic form, and further evolution of a solution after the change of type can be expected to lead to a singularity.

In fact, the mathematical singularity often signifies a real, physical event whose manifestation is an intense, highly localized deformation. Then the singularity may merely signify that insufficient physical characteristics have been modeled. If that is the case, inclusion of the appropriate extra physics will remove the singularity, if not all of the intense localization. In mathematical terms the equations are then said to have been regularized. That is exactly the case with adiabatic shear bands. If heat conduction and strain rate sensitivity are ignored, and if thermal softening dominates the material response, then the resulting equations can be seen to experience a change of type.

As an example, described by Wright (1987a), consider a rigid-plastic material that is a nonconductor of heat, that is insensitive to strain rate, with total conversion of plastic work to heat, and whose rate of work hardening is proportional to the rate of plastic work. We have

$$\begin{aligned}
\rho v_t &= s_y, & &\text{momentum;} \\
\rho c T_t &= s \dot{\gamma}^P, & &\text{energy;} \\
v_y &= \dot{\gamma}^P, & &\text{elasticity;} \\
s &= f(\kappa, T), & &\text{flow law;} \\
\kappa_t &= M(\kappa, T) s \dot{\gamma}^P, & &\text{work hardening.}
\end{aligned} \tag{5.21}$$

Combining the second and fifth equations leads to an ODE:

$$\frac{d\kappa}{dT} = \rho c M(\kappa, T). \tag{5.22}$$

With the solution of (5.22), $\kappa = \hat{\kappa}(T)$, inserted in the flow law, the evolving stress may be expressed as a function solely of temperature, $s = f[\hat{\kappa}(T), T] = \hat{S}(T)$. Flow stress generally increases with an increase in the work hardening parameter but decreases with increasing temperature. Therefore, it may be expected that the evolving flow stress has a maximum at $T = T_m$ where $(d\hat{S}/dT) = (\partial f/\partial \kappa)(d\hat{\kappa}/dT) + (\partial f/\partial T) = 0$. After the maximum, the slope of $\hat{S}(T)$ becomes negative. The space–time behavior of the temperature is now governed by a PDE that may be found by differentiating the balance of momentum, $s_{yy} = \rho v_{yt}$, and inserting the evolving stress where possible.

$$\frac{\partial^2 \hat{S}(T)}{\partial y^2} = \rho \frac{\partial}{\partial t}\left[\frac{\rho c \partial T/\partial t}{\hat{S}(T)}\right] \quad \text{or}$$

$$\frac{d\hat{S}}{dT}\frac{\partial^2 T}{\partial y^2} - \frac{\rho^2 c}{\hat{S}}\frac{\partial^2 T}{\partial t^2} + \frac{d^2\hat{S}}{dT^2}\left(\frac{\partial T}{\partial y}\right)^2 + \frac{\rho^2 c}{\hat{S}^2}\frac{d\hat{S}}{dT}\left(\frac{\partial T}{\partial t}\right)^2 = 0. \tag{5.23}$$

The flow stress is positive, but its slope, $d\hat{S}/dT$, may be positive or negative. At the maximum flow stress the slope changes sign, and the equation changes type from hyperbolic to elliptic. Further evolution may be expected to produce singularities in the solution.

Equation (5.23) would be regularized if it contained terms with higher than second derivatives, whose character was such that solutions no longer tended to a singularity but continued to evolve indefinitely. Such terms could have very small coefficients but still regularize the equation. In schematic fashion the equation might have two parts,

$$\varepsilon \mathcal{M}(z) + \mathcal{N}(z) = 0, \tag{5.24}$$

where z is the set of dependent variables, $\mathcal{N}(\cdot)$ is a nonlinear operator that "changes type" similar to (5.23), $\mathcal{M}(\cdot)$ is a regularizing operator with derivatives of higher order than $\mathcal{N}(\cdot)$, and ε is a small parameter (or parameters) with some definite physical meaning. In the limit that small parameters vanish, (5.24) reduces to an equation similar to (5.23) that changes type, but for any finite ε, the equation is "regularized," and solutions evolve indefinitely. Because of the singular perturbation provided by the first term in the equation, the regularized solution can be expected to show only strong localization, rather than an outright singularity.

The process works roughly as follows. In regions where the solution is not varying rapidly, the first term in (5.24) would be inconsequential compared with the second because of the small parameter. However, when strong gradients develop, as in the evolution toward a singularity, the product of the small parameter and the large gradient may make the first term comparable with the second so that a balance is achieved and regularization occurs. Equations that behave as described here have been called "equations with embedded change of type" by Varley (personnal communication). The equations that describe the formation of adiabatic shear bands represent a good example of this type of behavior.

5.4 Typical Numerical Results

The remainder of this chapter is devoted to reviewing numerical results from papers that explore some of the fundamental characteristics of adiabatic shear bands. Many researches prior to 1987 gave numerical results that faithfully followed the process of localization up to a point where resolution was lost as a result of insufficient grid refinement. A paper by Merzer (1982) was perhaps the first that showed the beginning of stress collapse, but again grid resolution was not sufficient to follow the process all the way to a fully formed band. However, he did comment that "substantial band weakening is associated with narrowing of the high strain-rate region," (p. 336) which may be taken to indicate the beginning of the fully formed stage of the shear band. Later, with a different model of viscoplasticity, Wright and Batra (1985) saw the same sort of behavior, and they remarked that "it is during the period of most rapid change that the shear band takes recognizable shape." (p. 211)

Following Einstein's aphorism that everything should be made as simple as possible, but no simpler and using a minimal version of viscoplasticity, Wright and Walter (1987) were able for the first time to follow the evolution of an adiabatic shear band from the initial, nearly homogeneous stage to the final, fully localized and slowly varying stage. Although the model was extremely simple, it displayed nearly all of the essential features that are characteristic of the localization process. Furthermore, when pertinent nondimensional parameters were varied, other important characteristics of localization were displayed.

The material was assumed to be rigid–perfectly plastic in isothermal deformation, to soften linearly with temperature, and to convert all plastic work to heat ($\beta = 1$). With these assumptions Equations (5.1) reduce to

$$
\begin{aligned}
\rho v_t &= s_y, & &\text{momemtum;} \\
\rho c \theta_t &= k\theta_{yy} + s v_y, & &\text{energy;} \\
s &= \kappa(1 - a\theta)(1 + b v_y)^m, & &\text{flow law.}
\end{aligned}
\tag{5.25}
$$

The material parameters ($\rho, c, k, \kappa, a, b, m$) are all assumed to be constants, and θ is the temperature relative to an arbitrary reference temperature.

For an experiment with a Kolsky torsion bar to be simulated the velocities were selected to be equal and opposite constants on the top and bottom surfaces of a slab of thickness $2H$ (so as to maintain a constant nominal strain rate, $\dot{\gamma}_0$), and the boundaries were assumed to be insulated. In order to simulate an arbitrary imperfection in the middle of the gage section, the initial temperature was assumed to be nearly constant, but with a slight symmetric increase in the center of the slab. The initial stress was given by $s = \kappa(1 + b\dot{\gamma}_0)^m$, and the initial strain rate was chosen to be compatible with the flow law so that $v_y \approx \dot{\gamma}_0$. The actual numerical values used in the calculation are not important but were chosen to represent a typical steel.

The solution was found by the method of lines. That is to say, after standard methods for finite elements are first used so as to discretize the equations in the y direction, the resulting system of coupled, first-order ODEs in time can be solved by any standard solver for a stiff set of nonlinear ODEs. To achieve the necessary spatial resolution, however, the grid was divided into more or less equal segments on a logarithmic scale. Details may be found in the original paper.

Typical solution surfaces for temperature, strain rate, and stress are shown in Figures 5.2, 5.3 and 5.4. The coordinates of the independent variables

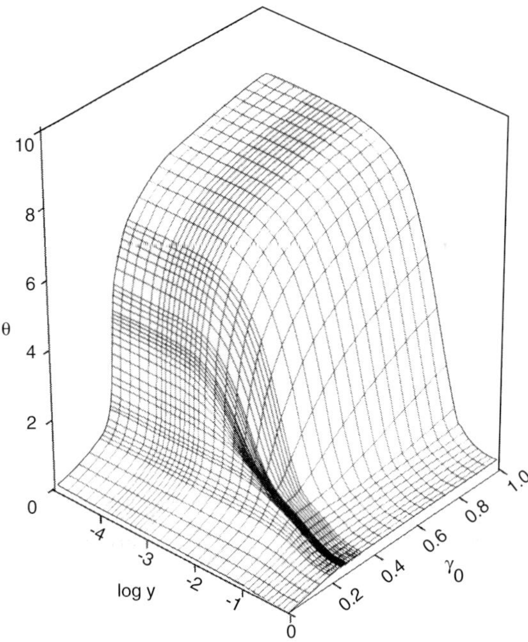

Figure 5.2. Surface of nondimensional temperature, $\theta = \rho c(T - T_0)/\kappa_0$, for a specimen in simple shear. There is a sharp rise in temperature when full localization occurs. Note the logarithmic length scale, measured from the center of the shear band. Compare the width of the pattern with that in Figure 5.3. (Reproduced from Wright and Walter 1987, with permission from Elsevier Science.)

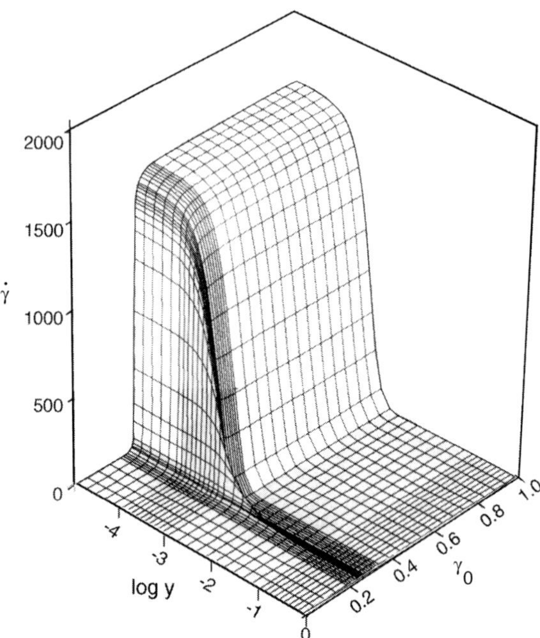

Figure 5.3. Surface of nondimensional strain rate, $\dot{\gamma} = v_y/\dot{\gamma}_0$, for a specimen in simple shear. There is a sharp rise in strain rate when full localization occurs. Note the logarithmic length scale, measured from the center of the shear band. Compare the width of the pattern with that in Figure 5.2. (Reproduced from Wright and Walter 1987, with permission from Elsevier Science.)

(space and time) have been nondimensionalized so that $Y = y/H$ and $\gamma_0 = \dot{\gamma}_0 t$. Similarly, the three response functions have been replaced by the nondimensional quantities $\Theta = \rho c\theta/\kappa$, $\dot{\gamma} = v_y/\dot{\gamma}_0$, and $\tau = s/k$. In these expressions $\dot{\gamma}_0$ is the nominal shearing strain rate for the slab.

In the plot of temperature, Figure 5.2, the initial perturbation is just visible along the spatial coordinate at the initial time where it occupies something less than the middle third of the slab. Although there is also an initial perturbation in strain rate, Figure 5.3, it is not visible on the scale of the plot. At first the perturbation grows slowly, but as the nominal strain approaches 0.27 or so, the strain rate is predicted to jump by a factor of approximately 1700, and the central nondimensional temperature to rise from a value of 0.1 to more than 8.0, a factor greater than 80. At the same time the stress, Figure 5.4, falls by a factor of three or four. Put in dimensional terms for a steel (heat capacity nearly 0.5×10^3 J/kg °C, density nearly 8×10^3 kg/m^3, and static shear strength of $\sim$0.6 GPa) with an initial temperature perturbation of approximately 15 °C, the strain rate starts at 5.0×10^2 s^{-1} and jumps to approximately 8.5×10^5 s^{-1}

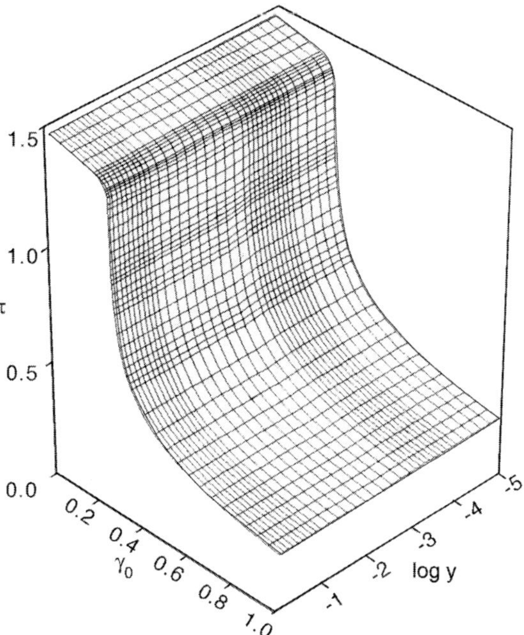

Figure 5.4. Surface of nondimensional stress, $\tau = s/\kappa_0$, for a specimen in simple shear. There is a sharp drop in stress when full localization occurs. Note the spatial uniformity. (Reproduced from Wright and Walter 1987, with permission from Elsevier Science.)

in approximately 20 μs. Over a somewhat longer time the temperature in the band increases by approximately 1200 °C while the shear stress drops from approximately 3/4 GPa to less than 1/4 GPa.

These are dramatic changes. Some other features of localization, also clearly visible in the three figures, are the distinctive profiles of temperature and strain rate that occur after severe localization. The thermal profile, although it occupies only ∼10% of the slab, is wider than the one for strain rate by more than an order of magnitude. Furthermore, the strain rate rapidly develops a steady profile, whereas the temperature slowly continues to grow and the profile to widen. In a qualitative sense all of these features are characteristic of band formation.

Other features of shear localization, equally as dramatic as those described above, become apparent when physical or kinematic parameters are varied. Figures 5.5, 5.6, and 5.7, also taken from Wright and Walter (1987), show how the delay to localization, the rise time at severe localization, and the final profile for strain rate all change as the nominal strain rate is changed.

Figure 5.5 shows the nominal strain at which the rapid localization occurs as a function of nominal strain rate in the present case. At a low strain rate the curve

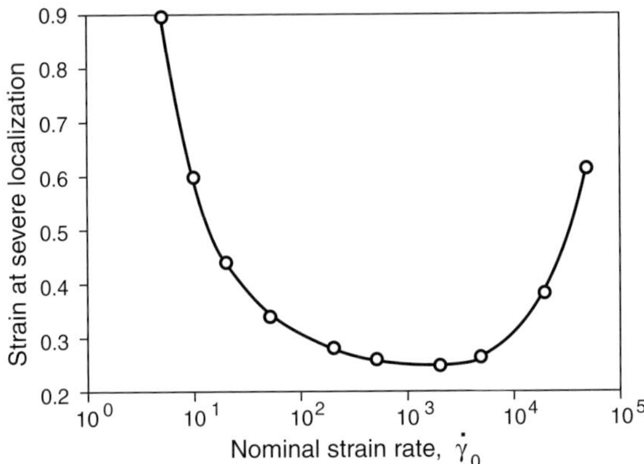

Figure 5.5. Strain at severe localization for a given defect in temperature at various applied strain rates. Heat conduction retards localization at low strain rates, and inertia retards it at high rates. Circles indicate points for calculation. Nominal gage length is 2.5 mm in a steel specimen. (Reproduced from Wright and Walter 1987, with permission from Elsevier Science.)

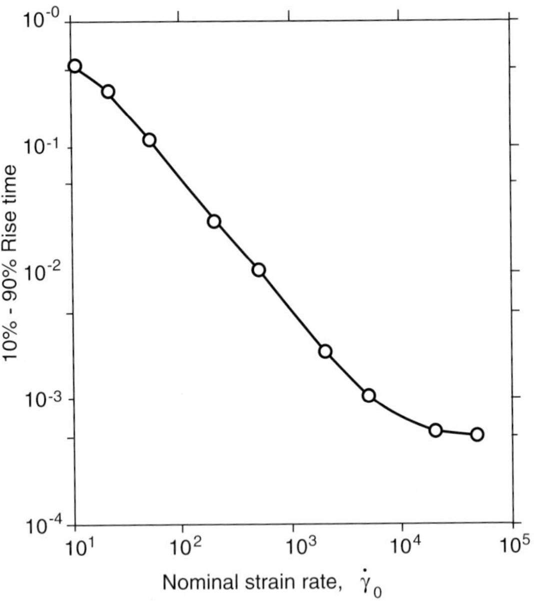

Figure 5.6. Rise time of severe localization for a specimen in simple shear. Nondimensional thermal diffusivity, $k = \bar{k}/\bar{\rho} c \dot{\gamma}_0 h^2$, decreases with applied strain rate, $\dot{\gamma}_0$, causing a shorter rise time. (Reproduced from Wright and Walter 1987, with permission from Elsevier Science.)

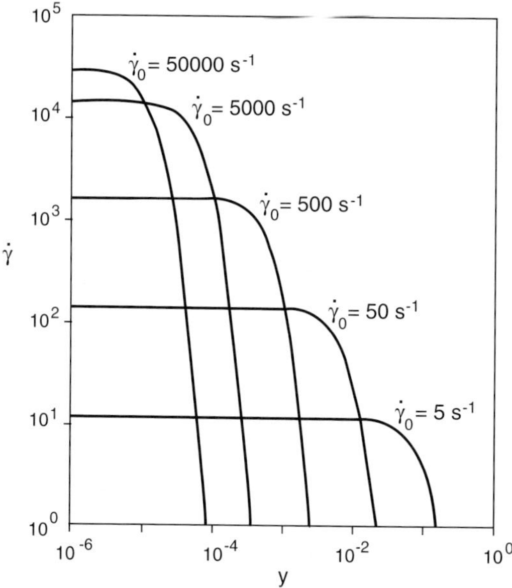

Figure 5.7. Late time profiles of strain rate when the shear band is fully formed, shown for applied strain rates varying from 5 to 50,000 s^{-1}. (Reproduced from Wright and Walter 1987, with permission from Elsevier Science.)

gives the appearance of tending toward infinity. This is a result of finite thermal conductivity, which smoothes the perturbation and overcomes the tendency to localize if the strain rate is small enough. As the nominal rate increases, heat production from plastic work in the energy equation of (5.25) becomes too rapid to be dissipated effectively by heat conduction, so localization tends to occur at smaller and smaller nominal strains. At higher strain rates thermal conductivity is wholly ineffective in retarding localization, but inertia eventually becomes strong enough to produce its own retarding effect. Then the curve has a minimum and the nominal strain at localization again increases rapidly. Similar results have also been found for work hardening materials by Klepaczko, Lipinski, and Molinari (1988) and in unpublished work by Walter, as reported by Wright (1994).

Figure 5.6 shows how the rise time of intense localization decreases as the nominal strain rate increases, and Figure 5.7 shows the late time profile of strain rate in the fully formed shear band. (The cross sections were all taken at the same nominal strain, so the profile for 50,000 s^{-1} may not actually be fully formed.) The value of the maximum strain rate increases by approximately an order of magnitude for each decade of increase in strain rate. Similarly, the width decreases by approximately an order of magnitude.

Virtually all of the principal characteristics of adiabatic shear bands, except the effects of work hardening, are illustrated at least qualitatively in Figures 5.2–5.7. A more realistic flow model for viscoplasticity and the inclusion both of thermoelastic effects and of the temperature dependence of the various coefficients would produce a quantitatively more precise description, but no significant qualitatively new features appear. In the next two chapters various problems are posed and analyzed, usually by asymptotic means, with the intent of finding scaling laws where possible and of shedding additional light on the process of localization.

6

One-dimensional problems, part II: Linearization and growth of perturbations

Because the equations used to describe one-dimensional shearing and the formation of adiabatic shear bands are fully coupled and highly nonlinear, it is not easy to analyze their behavior over a range of constitutive, boundary, and initial conditions. Numerical solutions are helpful, of course, but because they generally produce only one special case at a time, it may take an enormous amount of parametric study to cover the full range of possible behaviors. Even then it may be extremely difficult to extract patterns or simple universal relationships that effectively describe the possible responses in all their complexity. For that reason it is useful to develop a variety of approximate solutions, solutions to linearized equations, asymptotic solutions, or exact solutions to special cases. After a catalog of partial results is built, the generally expected behavior will begin to emerge. Furthermore, as special solutions are developed, characteristic groups of parameters will emerge again and again, and thus they will appear to have special significance.

For engineering applications it is always useful to have simple formulas available that characterize major features of the response. For adiabatic shear bands these might include various measures of the deformation pattern such as width, spacing, intensity, location, or other morphological features. They may also include dynamic features such as timing, rate of transition to the fully localized configuration, or various characterizations of propagating shear bands. In some cases such formulas may be shown to exist in principle, but the exact functional form may best be found experimentally or computationally. In other cases explicit algebraic formulas may be found, or at least suggested, from the solution of specific problems that are thought to be canonical in some sense. In their most useful form, such formulas depend only on well-established physical properties (density, initial yield strength, heat capacity, thermal conductivity, etc.)

and on physical dimensions or kinematic quantities (gage length, nominal strain rate, etc.).

These characterizing formulas always depend on nondimensional groupings of properties or *scaling parameters*. The formulas themselves may be called *scaling laws*. In this and the following chapters, special emphasis is laid on establishing useful scaling parameters and scaling laws for adiabatic shearing.

In this chapter linearized perturbation equations will be used to examine the behavior of small departures from homogeneous solutions. Not only are linear equations often easier to solve than nonlinear ones, but in addition they usually indicate important nondimensional groupings of physical, kinematic, and length parameters. Solutions to linear equations can also identify the initiation of patterns that persist in the fully nonlinear solutions.

6.1 Perturbations From Homogeneous Solutions: Linearized Equations

The notion of homogeneous solutions (i.e., solutions for stress, temperature, strain rate, etc. that depend only on time) was introduced in the last chapter. A standard way to test these solutions for stability is to introduce a small perturbation in addition to the homogeneous solution, to linearize the equations about the homogeneous solution, and then finally to examine the evolution of the perturbation with time. Roughly speaking, if arbitrary, small perturbations tend to decay rapidly with time, the homogeneous solution is stable, but if the perturbations tend to grow rapidly with time, then the homogeneous solution is unstable. The perturbation in effect acts as a probe for testing stability. However, as we shall see, analyzing and interpreting the behavior of the perturbation is not entirely straightforward and often requires some subtlety.

6.1.1 Frozen Coefficients

If the homogeneous solution were itself actually independent of space and time, as in the case of an elastic column under a compressive dead load, then the perturbation equations would constitute a system of linear equations with constant coefficients and homogeneous (or periodic) boundary conditions. Often it is reasonable to represent any perturbation as a Fourier series with time-dependent coefficients that are simple exponentials in time, where finding the form of the exponential reduces to finding the roots of a polynomial. Then if any of these roots have positive, real parts, the corresponding Fourier component will grow exponentially with time. Thus any arbitrary perturbation that contains that Fourier component will grow exponentially and may be deemed unstable.

When the homogeneous solution depends on time, the simple description above does not strictly apply. If the coefficients in the linearized equations are not constants, the time dependence of the solution is generally not just a simple exponential, so stability cannot be judged on the basis of the roots of a polynomial. Nevertheless, this approach has often been applied with the time-dependent coefficients frozen at arbitrary values of time. Justification is usually given by assuming that the rate of variation of the coefficients will be small compared with the time rate of an exponential. Although the idea of "frozen coefficients" has an intuitive appeal, there is little in the theory of ordinary differential equations to give it a sound foundation and much to make one wary.

In fact, numerous counterexamples have been given to show that the approximation of frozen coefficients may give completely erroneous indications of stability. The following case, cited by Hale (1980) in his book on ODEs, is from Markus and Yamabe (1960):

$$\begin{bmatrix} \dot{x} \\ \dot{y} \end{bmatrix} = \left\{ \begin{array}{cc} -1 + 3/2 \cos^2 t & 1 - 3/2 \cos t \sin t \\ -1 - 3/2 \cos t \sin t & -1 + 3/2 \sin^2 t \end{array} \right\} \begin{bmatrix} x \\ y \end{bmatrix}. \tag{6.1}$$

If the coefficients in the matrix are frozen at any value, say $t = t_0$, and the solution is assumed to have the form $x = x_0 e^{\lambda t}$, $y = y_0 e^{\lambda t}$, then λ must satisfy $\lambda^2 + 1/2\lambda + 1/2 = 0$, no matter what the value of t_0. The roots of the "characteristic" equation are $\lambda_{1,2} = 1/4[-1 \pm i\sqrt{7}]$. In particular, both roots have negative real parts, indicating that solutions "decay" with time. In fact, by direct computation it may be seen that one solution is given explicitly by

$$x = -e^{t/2}\cos t, \quad y = e^{t/2}\sin t, \tag{6.2}$$

which grows exponentially with time. Thus, the assumption of frozen coefficients in this case gives a false result.

As will be seen below, if there is work hardening in the fundamental equations for shear banding, the linearized equations generally show a rapid boundary layer in time after a perturbation is introduced, so again the instantaneous response is not indicative of the response over longer times. In other cases the method of frozen coefficients may give correct results, but there is no general theory to indicate when it applies. As a practical matter the approximation of frozen coefficients may be all that is available, but it is well to be aware of the dangers. Early works by Clifton (1980) and Bai (1982) both used the method and arrived at many conclusions that appear to be sound.

6.1.2 The Initial-Boundary Value Problem for Perturbations

An alternate approach is to pose a definite initial-boundary value problem, to maintain the time dependence of the coefficients, and to solve the problem in as general a form as possible. This is the approach taken by Wright (1992), and which will be described in some detail here. Accordingly, let us consider a slab of thickness $2H$ whose upper and lower surfaces are sheared with equal and opposite velocities, $\pm V$, so that the average or nominal strain rate is $\dot{\gamma}_0 = V/H$. The two surfaces are also fully insulated so that the temperature gradient vanishes on both boundaries. This problem is meant to approximate an experiment with the torsional split Hopkinson bar. The equations for a rigid-plastic material, repeated again, are

$$\bar{s}_{\bar{y}} = \bar{\rho}\,\bar{v}_{\bar{t}}, \qquad\qquad \text{momentum;}$$

$$\bar{\rho}c\bar{\theta}_{\bar{t}} = \bar{k}\bar{\theta}_{\bar{y}\bar{y}} + \bar{s}\bar{v}_{\bar{y}}, \qquad \text{energy;}$$

$$\bar{\kappa}_{\bar{t}} = \bar{M}(\bar{\kappa},\,\bar{\theta})\bar{s}\bar{v}_{\bar{y}}, \qquad \text{work hardening;}$$

$$\bar{s} = \bar{F}(\bar{\kappa},\,\bar{\theta},\,\bar{v}_{\bar{y}}), \qquad \text{flow law.}$$

$$(6.3)$$

The boundary conditions are $\bar{v}(\pm H, \bar{t}) = \pm V$, $\bar{\theta}_{\bar{y}}(\pm H, \bar{t}) = 0$. The overbars indicate dimensional quantities that will be replaced below by nondimensional ones. The plastic work is assumed to be totally converted to heat, the rate of work hardening has been assumed to be proportional to the rate of plastic work, as in (4.10), and the relative temperature has been used, that is, $\bar{\theta} = T - T_0$. The initial conditions for homogeneous shearing are $\bar{v}_{\bar{y}} = \dot{\gamma}_0$, $\bar{\theta} = 0$, $\bar{\kappa} = \kappa_0$, and $s_0 = \bar{F}(\kappa_0, 0, \dot{\gamma}_0)$. The flow stress $\bar{F}$ and the plastic modulus $\bar{M}$ have the following properties: $\bar{F} > 0$ for positive flow stress, $\bar{F}_{\bar{\kappa}} > 0$ for work hardening, $\bar{F}_{\bar{\theta}} < 0$ for thermal softening, $\bar{F}_{\dot{\gamma}} > 0$ for strain rate hardening, and $\bar{M} \geq 0$ for work hardening.

For a simple and straightforward scheme to nondimensionalize these equations, let

$$y = \bar{y}/H, \quad t = \dot{\gamma}_0 \bar{t}, \quad v = \bar{v}/H\dot{\gamma}_0,$$

$$s = \bar{s}/s_0, \quad \kappa = \bar{\kappa}/\kappa_0, \quad \theta = \bar{\rho}c\bar{\theta}/s_0, \quad v_y = \bar{v}_{\bar{y}}/\dot{\gamma}_0,$$

$$M(\kappa, \theta) = (s_0/\kappa_0)\bar{M}(\kappa_0\kappa, s_0\theta/\bar{\rho}c), \quad F(\kappa, \theta, v_y) = s_0^{-1}\bar{F}(\kappa_0\kappa, s_0\theta/\bar{\rho}c, \dot{\gamma}_0 v_y),$$

$$\rho = \bar{\rho}\dot{\gamma}_0^2 H^2/s_0, \quad k = \bar{k}/\bar{\rho}c\dot{\gamma}_0 H^2$$

$$(6.4)$$

The nondimensionalized equations appear almost the same as (6.3), but without overbars.

$$s_y = \rho v_t, \qquad\qquad \text{momentum;}$$

$$\theta_t = k\theta_{yy} + sv_y, \quad \text{energy;}$$

$$\kappa_t = M(\kappa, \theta)sv_y, \quad \text{work hardening;}$$

$$s = F(\kappa, \theta, v_y), \quad \text{flow law.}$$

(6.5)

To get some sense of the magnitude of various quantities, again consider a high-strength steel in the torsional experiment. Typical values of the dimensions and physical properties might be $H \approx 10^{-3}$ m, $\dot\gamma_0 \approx 10^3 \text{ s}^{-1}$, $s_0 \approx 8 \times 10^8$ Pa, $\bar\rho \approx 8000$ kg/m^3, $c \approx 500$ J/kg K, and $\bar k \approx 50$ W/mK. With these numbers $\rho \approx 10^{-5}$, $k \approx 10^{-2}$, and the temperature and velocity scales are $s_0/\bar\rho c \approx 200\,°$C and $V = H\dot\gamma_0 \approx 1$ m/s. The nondimensional density is so small that the right-hand side of the momentum equation may be set to zero. Solutions to these reduced equations may be termed quasi-static. Recall that the numerical results described in Chapter 5 show little spatial variation of stress, even when there is strong temporal variation. The reason for this feature is now apparent. However, if the boundary velocity, V, is substantially increased, the nondimensional density, ρ, may become large enough that the quasi-static approximation is no longer appropriate. If not all plastic work is converted to heat, $\beta = \text{const} < 1$ [see Equation (3.57) and the discussion following it], it can be absorbed into the scaling by setting $\theta = \rho c(T - T_0)/\beta s_0$. The effect on the scaling numbers that would be brought about by changes in the physical properties of the material may be calculated readily.

Two other important physical quantities that are characteristic of the flow law are the thermal sensitivity and the strain rate sensitivity. These are defined as

$$a = -\frac{\partial F}{\partial \theta} = -\frac{1}{\bar\rho c}\frac{\partial \bar F}{\partial \bar\theta}, \quad \text{thermal sensitivity;}$$

$$m = \frac{\partial \ln F}{\partial \ln v_y} = \frac{\partial \ln \bar F}{\partial \ln \bar v_{\bar y}}, \quad \text{strain rate sensitivity.}$$

(6.6)

At room temperature typical values for metals are $a = \mathcal{O}(10^{-1})$ and $m = \mathcal{O}(10^{-2})$.

Suppose that the initial conditions at a constant strain rate, $\dot\gamma_0$, are nearly constant, with average temperature T_0, strength κ_0, and stress $s_0 = \bar F(\kappa_0, 0, \dot\gamma_0)$. Because the nondimensional temperature is based on the deviation from average temperature, its initial average must vanish; that is, $\int_{-1}^{+1}\theta_0(y)\,dy = 0$. Similarly, the initial average of the nondimensional strength must be one; that is, $\int_{-1}^{+1}\kappa_0(y)\,dy = 1$. Initial values are given by $\theta_0(y) = \theta(y, 0)$, and so on.

With the momentum equation replaced by $s_y = 0$ for the quasi-static approximation ($\rho \ll 1$), let us now examine the behavior of perturbations $\tilde{s}$, $\tilde{\theta}$, $\tilde{\kappa}$, and $\tilde{v}$, which will be regarded as small compared with the homogeneous solution, $S(t)$, $\Theta(t)$, $K(t)$, and y.

$$s = S(t) + \tilde{s}(t),$$
$$\theta = \Theta(t) + \tilde{\theta}(y, t),$$
$$\kappa = K(t) + \tilde{\kappa}(y, t),$$
$$v = y + \tilde{v}(y, t).$$

$$(6.7)$$

The initial and boundary conditions for all functions are

$$S(0) = 1, \quad \Theta(0) = 0, \quad K(0) = 1,$$
$$\tilde{\theta}_0(y) = \tilde{\theta}(y, 0) = \theta_0(y), \quad \tilde{\kappa}_0(y) = \tilde{\kappa}(y, 0) = \kappa_0(y) - 1,$$
$$\tilde{v}(0, t) = \tilde{v}(1, t) = 0, \quad \tilde{\theta}_y(0, t) = \tilde{\theta}_y(1, t) = 0.$$

$$(6.8)$$

The functions $\theta_0(y)$ and $\kappa_0(y)$ are arbitrary functions with averages of zero and one, respectively, so that $\int_{-1}^{+1} \tilde{\theta}_0(y)\, dy = \int_{-1}^{+1} \tilde{\kappa}_0(y)\, dy = 0$. The initial values for $\tilde{v}(y, 0)$ must be chosen so as to be compatible with the perturbed flow law given in Equation (6.10).

The homogeneous solution must satisfy the quasi-static form of (6.5) by definition:

$$\dot{\Theta} = S,$$
$$\dot{K} = M(K, \Theta)S,$$
$$S = F(K, \Theta, 1).$$

$$(6.9)$$

Because $S > 0$, the average temperature increases, and because M is nonnegative, the work hardening parameter cannot decrease. In turn, the perturbations must satisfy

$$\begin{aligned}
&\tilde{s}_y = 0, &&\text{momentum;}\\
&\tilde{\theta}_t = k\tilde{\theta}_{yy} + S\tilde{v}_y + \tilde{s}, &&\text{energy;}\\
&\tilde{\kappa}_t = (M_\kappa\tilde{\kappa} + M_\theta\tilde{\theta})S + M(\tilde{s} + S\tilde{v}_y), &&\text{work hardening;}\\
&\tilde{s} = F_\kappa\tilde{\kappa} + F_\theta\tilde{\theta} + F_{\dot{\gamma}}\tilde{v}_y, &&\text{flow law.}
\end{aligned}$$

$$(6.10)$$

Here subscripts such as $M_\kappa = \partial M / \partial \kappa$ and so on denote derivatives with respect to the argument indicated. All functions in capital letters are evaluated on the homogeneous solution, and thus all are functions of time, for example, $S(t)$ and $M[K(t), \Theta(t)]$, and so on. As a consequence, although the system of Equations (6.10) is linear, all but the first equation have time-dependent coefficients, and so must be treated with care, as indicated by the example given in Equations (6.1) and (6.2).

It is now useful to introduce a new dependent variable,

$$\tilde{\lambda} = \tilde{\kappa} - M(t)\tilde{\theta}, \tag{6.11}$$

which is a composite measure of the perturbation. Note that $\tilde{\lambda} < 0$ when $\tilde{\kappa} < 0$ and $\tilde{\theta} > 0$. That is to say, either a decrease in strength or an increase in temperature tends to make $\tilde{\lambda}$ negative. Because the initial averages of both $\tilde{\kappa}$ and $\tilde{\theta}$ vanish by definition, the initial average of $\tilde{\lambda}$ vanishes, too. With the aid of Equations (6.9) and (6.10), $\tilde{\lambda}$ may be shown to satisfy

$$\tilde{\lambda}_t = M_\kappa S \tilde{\lambda} - kM\tilde{\theta}_{yy}. \tag{6.12}$$

With the boundary condition in (6.8), averaging (6.12) leads to $d/dt \int_{-1}^{+1} \tilde{\lambda} dy = M_\kappa S \int_{-1}^{+1} \tilde{\lambda} dy$, which has the solution

$$\int_{-1}^{+1} \tilde{\lambda}(y, t) \, dy = \left[\int_{-1}^{+1} \tilde{\lambda}_0(y) \, dy\right] \exp\left\{\int_0^t M_\kappa S dt\right\} = 0. \tag{6.13}$$

Because the initial average of $\tilde{\lambda}$ vanishes, its average vanishes for all times.

Because $\tilde{s}$ depends at most on t and not on y, averaging of the perturbed energy equation and the perturbed flow law leads to

$$\frac{d}{dt} \int_{-1}^{+1} \tilde{\theta} dy = 2\tilde{s} = 2(F_\kappa M + F_\theta) \int_{-1}^{+1} \tilde{\theta} dy, \tag{6.14}$$

where the definition of $\tilde{\lambda}$, the boundary condition on $\tilde{v}$, and (6.13) have been used. Again the result is the vanishing of an average:

$$\int_{-1}^{+1} \tilde{\theta}(y, t) \, dy = 0. \tag{6.15}$$

The most important consequence of (6.15) is that

$$\tilde{s} = 0, \tag{6.16}$$

which follows directly from (6.14), but in addition, from the definition of $\tilde{\lambda}$, the average of $\tilde{\kappa}$ must also vanish for all times. The results so far are that the spatially averaged values of perturbations of stress, temperature, strength, and strain rate must all vanish.

Before proceeding further, let us return briefly to the homogeneous equations (6.9) to note that the first two may be combined to yield $dK/d\Theta = M(K, \Theta)$ with solution $K = \hat{K}(\Theta)$. Then the third of (6.9) may be written $S = F[\hat{K}(\Theta), \Theta, 1] = \hat{S}(\Theta)$, and the combination of terms $F_\kappa M + F_\theta$, which appears repeatedly, has the significance $d\hat{S}/d\Theta = F_\kappa M + F_\theta$ evaluated on the homogeneous solution. Finally, if the dependence of flow stress on strain rate obeys a power law, as is often found experimentally over significant ranges of strain rate, the reduced flow stress might be written $s = F(\kappa, \theta, \dot{\gamma}) = \hat{F}(\kappa, \theta)\dot{\gamma}^m$. In this case the strain rate sensitivity is given by $m = \partial \ln s / \partial \ln \dot{\gamma} = (\dot{\gamma}/F)\partial F/\partial\dot{\gamma}$

and is to be evaluated on the homogeneous solution where the nondimensional strain rate is $\dot{\gamma} = 1$. In a similar way the homogeneous strain rate sensitivity for a more general material that obeys the work hardening and flow law of (6.5) may be written as

$$m(\Theta) = \frac{F_{\dot{\gamma}}[\hat{K}(\Theta), \Theta, 1]}{F[\hat{K}(\Theta), \Theta, 1]} = \frac{F_{\dot{\gamma}}(t)}{\hat{S}(t)}. \tag{6.17}$$

With these new definitions the energy equation in (6.10) and Equation (6.12), repeated here, define a coupled pair of linear PDEs for the perturbations in the quasi-static case.

$$\tilde{\theta}_t = k\tilde{\theta}_{yy} - \frac{\hat{S}_\theta}{m}\tilde{\theta} - \frac{F_\kappa}{m}\tilde{\lambda},$$

$$\tilde{\lambda}_t = M_\kappa \hat{S}\tilde{\lambda} - kM\tilde{\theta}_{yy}. \tag{6.18}$$

Boundary conditions remain as stated in (6.8).

6.2 Quasi-Static Solutions for Special Cases on a Finite Interval

The major characteristics of solutions to (6.18) may be found by examining a sequence of special cases with increasing complexity.

6.2.1 *Perfect Plasticity With Finite Thermal Conductivity*

Suppose that there is no work hardening so that $M \equiv 0$, and furthermore, suppose that the initial strength is constant so that $\tilde{\kappa}_0(y) = 0$. In this case, the dimensional flow law reads $\bar{s} = \kappa_0 \hat{F}(\bar{\theta}, \dot{\bar{\gamma}})$, where κ_0 is the initial quasi-static shear strength. From (6.9) it is apparent that $\Theta(t)$ is an increasing function, but that $\hat{S}(t)$ is a positive but decreasing function because it is always assumed that $F_\Theta < 0$. Then the solution to the second equation of (6.18) is simply

$$\tilde{\lambda}(y, t) = \tilde{\lambda}(y, 0) = \tilde{\kappa}(y, 0) - M(0)\tilde{\theta}(y, 0) \equiv 0. \tag{6.19}$$

Note, however, that even though $\tilde{\kappa}_0 = 0$, (6.19) does not imply that $\tilde{\theta}_0 = 0$ because $M(0) = 0$ in this case.

The reduced energy equation becomes a diffusion equation with a time-dependent source,

$$\tilde{\theta}_t = k\tilde{\theta}_{yy} - (\hat{S}_\theta/m)\tilde{\theta}, \tag{6.20}$$

which has the solution

$$\tilde{\theta} = \psi \exp\left\{-\int_0^t \frac{\hat{S}_\theta}{m}\,dt\right\},$$

$$\psi_t = k\psi_{yy},$$

$$\psi(y, 0) = \tilde{\theta}_0(y), \quad \psi_y(0, t) = \psi_y(1, t) = 0. \tag{6.21}$$

The behavior of the simple diffusion function, ψ, is to decay uniformly to its average, which is identically equal to zero. For example, if the initial temperature distribution is symmetric in y, ψ may be represented by a cosine series, $\psi = \sum_1^\infty \theta_n e^{-k(n\pi)^2 t} \cos n\pi y$, where the θ_n are the Fourier components of the initial temperature perturbation.

The exponential factor in (6.21) may be recast into a more informative form to show that it promotes growth, which is opposed to the diffusive decay. Because the terms in the integral refer to the homogeneous solution, (6.9) allows time, t, to be replaced by temperature, because $dt = d\theta/\hat{S}$. Then the exponential becomes

$$\exp\left\{-\int_0^t \frac{\hat{S}_\theta}{m} \, dt\right\} = \exp\left\{-\int_0^{\Theta(t)} \frac{\hat{S}_\theta}{m\hat{S}} \, d\theta\right\}. \tag{6.22}$$

If m is constant, the value of the exponential is $\hat{S}^{-1/m}$, and because $\hat{S}$ is a function that decreases from one as a result of thermal softening, and $m \ll 1$, the exponential multiplier in (6.21) rapidly provides powerful amplification to offset the diffusion. As a simple example, suppose $\hat{S} = e^{-a\Theta}$. Then it is easy to show that $a\Theta = \ln(1 + at)$, $\hat{S} = 1/(1 + at)$, and (6.22) works out to be $(1 + at)^{1/m}$. Figure 6.1 illustrates the effect with Θ rather than t as the independent variable.

As a practical matter, because the perturbation solution is of interest only for early times (i.e., small nominal strains, because $t = \dot{\gamma}_0 \bar{t} = \gamma$), the exponential

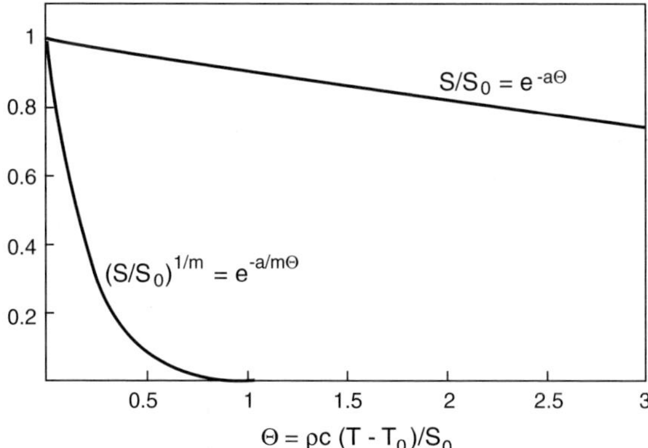

Figure 6.1. Strain rate sensitivity and thermal softening have a powerful influence on amplification of perturbations. The upper curve in the figure shows the effect of thermal softening; the lower curve is the reciprocal of the amplifying factor in Equation (6.21).

may be approximated by replacing both $\hat{S}_\theta$ and m by their initial values. Thus,

$$\exp\left\{-\int_0^t \frac{\hat{S}_\theta}{m}\,dt\right\} \sim e^{at/m} = \exp\left\{\frac{s_0\bar{a}}{\bar{\rho}\,cm}\gamma\right\}. \tag{6.23}$$

For linear softening, $\hat{S} = 1 - a\Theta$, the exponential $e^{at/m}$ is exact. When (6.23) is combined with the Fourier form for ψ, it is apparent that those Fourier components for which $a/m < k(n\pi)^2$ will decay and only those for which $a/m > k(n\pi)^2$ will tend to grow initially. Evidently all higher components decay, and only a finite number of the lowest-order components can grow. In particular, no Fourier component can begin to grow if $a/m < k\pi^2$. In dimensional terms, this says that for steady shearing of a finite strip of material that has no work hardening, no adiabatic shear band can form if

$$-\frac{\partial}{\partial\bar{\theta}}\hat{F}(0,\dot{\gamma}_0) < \frac{\bar{k}m\pi^2}{\kappa_0\dot{\gamma}_0 H^2}. \tag{6.24}$$

A version of this formula was first given by Wright and Walter (1987). Note that both density and heat capacity cancel out of the inequality when it is put into dimensional terms. To get some sense of the implications that (6.24) has for real materials, it is useful to insert order-of-magnitude quantities into the equation. For example, for a typical strong material with the left-hand side estimated to be $\mathcal{O}(10^{-3})\,°\mathrm{C}^{-1}$ (i.e., the tangential projection of material strength goes from κ_0 to zero in approximately $10^3\,°\mathrm{C}$), $\bar{k} = \mathcal{O}(10^2)\mathrm{W/m}\,°\mathrm{C}$, $m = \mathcal{O}(10^{-2})$, $\kappa_0 = \mathcal{O}(10^9)\,\mathrm{Pa}$, and $H = \mathcal{O}(10^{-3})\,\mathrm{m}$, then strain rates greater than $\dot{\gamma}_0 = \mathcal{O}(10)\,\mathrm{s}^{-1}$ will tend to produce adiabatic shear bands. With variations in the physical properties and dimensions for different materials, it is easy to see that the minimum strain rate to produce shear bands could readily vary either up or down by a factor of five or more. Nevertheless, it seems clear that shear bands often may form at relatively low strain rates.

The formula clearly indicates that thermal conductivity and strain rate sensitivity are stabilizing material properties, but thermal softening and strength are destabilizing. Higher nominal strain rate and greater gage length are also destabilizing. Finally, shear bands may be expected to begin to form at strain rates of only a few tens per second.

6.2.2 Work Hardening Without Thermal Conductivity: the Early Response

The deformation is adiabatic in this case, but there is work hardening. Again the homogeneous temperature increases, but the adiabatic flow stress typically increases at first, reaches a maximum, and finally decreases monotonically. Equation (6.18) becomes

$$\tilde{\theta}_t = -\frac{\hat{S}_\theta}{m}\tilde{\theta} - \frac{F_\kappa}{m}\tilde{\lambda},$$

$$\tilde{\lambda}_t = M_\kappa S \tilde{\lambda}. \tag{6.25}$$

The second equation has become uncoupled from the first and has the immediate solution

$$\tilde{\lambda}(y, t) = \{\tilde{\kappa}_0(y) - M(0)\tilde{\theta}_0(y)\}\exp\left\{\int_0^t M_\kappa(t')S(t')\,dt'\right\}. \tag{6.26}$$

Because $\tilde{\lambda}$ is now known, the first equation can be integrated explicitly, as well:

$$\tilde{\theta}(y, t) = e^{-\int_0^t (S_\theta/m)\,dt'}\left\{\tilde{\theta}_0(y) - \int_0^t \frac{F_\kappa}{m}\tilde{\lambda}(y, t')e^{\int_0^{t'} (S_\theta/m)\,dt''}\,dt'\right\} \tag{6.27}$$

Some sense of the implications of Equations (6.26) and (6.27) can be found by examining further specializations. If the work hardening modulus, M, is independent of temperature and depends only on κ as in Equation (4.10), for example, according to (6.4) $M(\kappa) = (ns_0/\gamma_0\kappa_0)\kappa^{-1/n}$ in nondimensional terms, then the exponential in (6.26) may be evaluated explicitly. According to (6.9) in this case, integration of the homogeneous solution on time may be replaced by κ because $Sdt = d\kappa/M$. Thus, $e^{\int_0^t M_\kappa(t')S(t')\,dt'} = M(t)/M(0)$, so that

$$\tilde{\lambda}(y, t) = \frac{M(t)}{M(0)}\tilde{\lambda}(y, 0) = \frac{M(t)\tilde{\kappa}_0(y)}{M(0)} - M(t)\tilde{\theta}_0(y). \tag{6.28}$$

If (6.4) is used to generate the plastic modulus, as suggested above, then $M(0) = ns_0/\gamma_0\kappa_0$ and $M(t)/M(0) = \kappa^{-1/n}$. Because κ increases from one as work hardening progresses, and n is generally one or smaller; the plastic modulus decreases from its initial value. In addition, because the plastic modulus is positive, the composite defect does not change sign. Thus, in the absence of heat conduction, points that initially have a positive (negative) defect remain positive (negative).

Equation (6.27) may also be further reduced and simplified if the strain rate sensitivity is constant. Then as in (6.22), $\exp\{-\int_0^t (\hat{S}_\theta/m)\,dt'\} = \hat{S}^{-1/m}$. In this case however, $\hat{S}$ increases from one initially, so the exponential rapidly decreases, rather than increases as previously. Equation (6.27) may now be written as

$$\tilde{\theta}(y, t) = \tilde{\theta}(y, 0)\hat{S}^{-1/m}(t) - \frac{\tilde{\lambda}(y, 0)}{mM(0)}\int_0^t F_\kappa M\left[\frac{\hat{S}(t')}{\hat{S}(t)}\right]^{1/m}\,dt'. \tag{6.29}$$

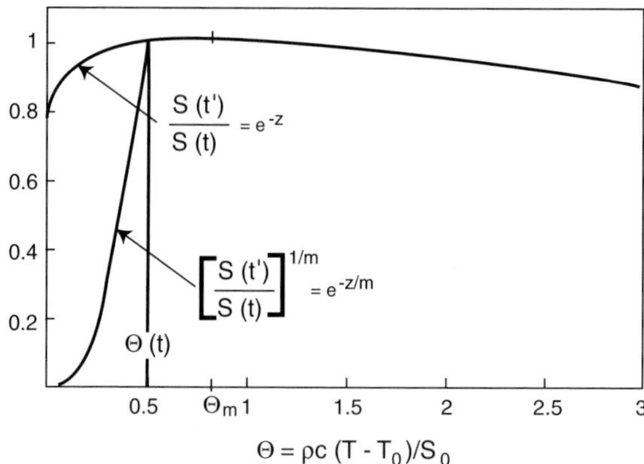

Figure 6.2. The homogeneous response is shown in the upper curve, normalized at the upper limit of integration in Equation (6.29). The stress ratio shown in brackets in (6.29), when raised to a large power, is a rapidly varying function, as shown by the lower curve in the figure. The asymptotic approximation takes advantage of this fact.

An asymptotic evaluation of the integral (e.g., see Jeffreys 1962 or Bender and Orszag 1978) provides further reduction. Recall that in a typical case, the homogeneous stress first increases, then reaches a maximum, and finally decreases, as in Figure 6.2. On the rising part of the curve, we have $0 \le t' \le t < t_m$. Consequently, $\hat{S}(t')/\hat{S}(t) < 1$, and the large exponent, $1/m$, magnifies the contrast. Thus, the ratio of stresses is one at the upper limit of the integral, and $[\hat{S}(t')/\hat{S}(t)]^{1/m}$ rapidly tends toward zero away from the upper limit, as shown in the figure. As a result only the values of the integrand near the upper limit contribute substantially to the integral. This effect can be captured quantitatively by changing the variable of integration. Thus, we define z as a function of t' by letting $e^{-z} = \hat{S}(t')/\hat{S}(t)$, so that when $t' = 0$, we have $z = z(0) = \ln \hat{S}(t) > 0$, and when $t' = t$, we have $z = 0$. With this new variable, the integral becomes $\int_{z(0)}^{0} e^{-z/m} F_\kappa M (dt'/dz) \, dz$ exactly.

On the interval of integration the exponential has its maximum at $z = 0$, and because m is small, the exponential is rapidly varying. However, if the terms that follow the exponential are only slowly varying by comparison, they may be expanded in a Taylor series about $z = 0$, and the integral may be evaluated explicitly term by term. After the expansion, the integral becomes

$$\int_{z_0}^{0} e^{-z/m}(a_0 + a_1 z + \cdots) \, dz = -ma_0\left(1 - e^{-z_0/m}\right)$$

$$+ a_1 m\left[z_0 e^{-z_0/m} - m\left(1 - e^{-z_0/m}\right)\right] + \mathcal{O}(m^3).$$

$$(6.30)$$

Because $z_0 e^{-z_0/m}$ has a maximum of me^{-1}, the second term is $\mathcal{O}(m^2)$, and the series is asymptotic in the small parameter, m. The leading coefficient, a_0, is just $F_\kappa M(dt'/dz)$ evaluated at $t' = t$.

The derivative, dt'/dz, is found by differentiating the definition of z. Thus, from $\hat{S}(t') = \hat{S}(t)e^{-z}$, we obtain $\hat{S}_\Theta(d\Theta/dt')(dt'/dz) = -\hat{S}(t)e^{-z}$, where the left-hand side is evaluated at t' in general. However, from the homogeneous solution we have $d\Theta/dt' = \hat{S}(t')$, so when the differentiated expression is evaluated at $t' = t$ (i.e., $z = 0$), we find that $dt'/dz = -1/\hat{S}_\Theta(t)$. Thus, the $\mathcal{O}(1)$ term in the asymptotic evaluation of (6.29) is

$$\tilde{\theta}(y, t) \sim \tilde{\theta}_0(y)\hat{S}^{-1/m}(t) - \tilde{\lambda}_0(y)\frac{M(t)F_\kappa(t)}{M(0)\hat{S}_\Theta(t)}\left[1 - \hat{S}^{-1/m}(t)\right] + \mathcal{O}(m). \quad (6.31)$$

From the definition of $\tilde{\lambda}$ in (6.11) together with (6.31), the perturbation of strength is found to be

$$\tilde{\kappa}(y, t) \sim \tilde{\kappa}_0(y)\frac{M(t)}{M(0)} - M(t)\left[\tilde{\theta}_0(y) + \tilde{\lambda}(y, t)\frac{F_\kappa(t)}{\hat{S}_\Theta(t)}\right]\left[1 - \hat{S}^{-1/m}(t)\right] + \mathcal{O}(m).$$

$$(6.32)$$

The perturbed strain rate now follows algebraically from the perturbed flow law in (6.10), the representations for $\tilde{\kappa}$ and $\tilde{\theta}$, and the previous result that $\tilde{s} = 0$.

$$\tilde{v}_y = -\frac{1}{m}\hat{S}^{-(1+m)/m}(t)\left[F_\theta(t)\tilde{\theta}_0(y) + \frac{M(t)}{M(0)}F_\kappa(t)\tilde{\kappa}_0(y)\right] + \mathcal{O}(1). \quad (6.33)$$

These last three formulas are notable in several respects: they contain an initial boundary layer, they show a simplified structure after the boundary layer, and they indicate a general growth of the perturbations even though the homogeneous stress is still rising.

First, there is a boundary layer in time at $t = 0$ because $\hat{S}^{-1/m}$ is a function that starts at a value of one and then decays rapidly toward zero. The boundary layer vanishes only if the total coefficient of $\hat{S}^{-1/m}$ vanishes at $t = 0$. When it is recalled that $\hat{S}_\Theta = F_\kappa M + F_\theta$ and that $\tilde{\lambda} = \tilde{\kappa} - M\tilde{\theta}$ by definition, the condition that the boundary layer vanish for all three equations is $\hat{S}_\Theta(0)\tilde{\theta}_0(y) + F_\kappa(0)\tilde{\lambda}_0(y) = 0$, which may also be written as

$$F_\kappa(0)\tilde{\kappa}_0(y) + F_\theta(0)\tilde{\theta}_0(y) = 0. \quad (6.34)$$

After taking account of the flow law in (6.10) and the fact that $\tilde{s} = 0$, Equation (6.34) also implies directly that $\tilde{v}_y(y, 0) = 0$. This indicates that the $\mathcal{O}(1)$ term in (6.33), which has not been worked out in detail, must actually vanish at $t = 0$. In other words, after the boundary layer the perturbation in strain rate must be small.

Second, note that after the boundary layer the solution tends toward the simple solution

$$\tilde{\theta}(y, t) \sim - \tilde{\lambda}(y, t)\frac{F_\kappa(t)}{\hat{S}_\Theta(t)} + \mathcal{O}(m),$$

$$\tilde{\kappa}(y, t) \sim \tilde{\lambda}(y, t)\frac{F_\theta(t)}{\hat{S}_\Theta(t)} + \mathcal{O}(m), \tag{6.35}$$

$$\tilde{v}_y(y, t) \sim \mathcal{O}(1),$$

where $\tilde{\lambda}$ is given by (6.28), and the $\mathcal{O}(1)$ term in the strain rate actually vanishes at $t = 0$. Notice that the postboundary layer solution for temperature could have been found directly from (6.25) given the assumption that $m \ll 1$. In any event, by substitution it may be seen that Equation (6.35) is fully consistent with the definition $\tilde{\lambda} = \tilde{\kappa} - M\tilde{\theta}$. Also, by cross multiplication and addition of the first two expressions, it may be seen that the structure after the boundary layer is the same as the initial structure that suppresses the boundary layer altogether, to within terms of order m:

$$F_\kappa(t)\tilde{\kappa}(y, t) + F_\theta(t)\tilde{\theta}(y, t) = \mathcal{O}(m). \tag{6.36}$$

In effect then, even though the reduced Equations (6.18) indicate that there will be two independent solutions because the system is second order in time, there is really only one solution that matters, and that is the one that survives after the boundary layer.

Finally, notice that at a point where $\tilde{\lambda}$, the composite defect, is negative initially (that is to say $\tilde{\kappa}_0 < 0$ and/or $\tilde{\theta}_0 > 0$), Equation (6.35) indicates that the perturbations in strength and temperature will both be positive after the boundary layer because of thermal softening, $F_\theta < 0$. This is an indication that the material at the core of a shear band will experience both increased work hardening and increased temperature relative to the homogeneous case, and that the differentials over neighboring points begin to form early in the process. Conversely, at points where $\tilde{\lambda}_0 > 0$ ($\tilde{\kappa}_0 > 0$ and/or $\tilde{\theta}_0 < 0$), perturbations in both strength and temperature become negative after the initial boundary layer.

On the rising portion of the homogeneous stress–strain curve, it seems reasonable to conclude that the initial rapid boundary layer in the perturbed solution is meaningless as far as judging whether or not that part of the homogeneous response is stable or not. In fact, Equation (6.35) indicates that after the boundary layer, as the homogeneous slope, $\hat{S}_\theta$, decreases toward zero, the perturbed temperature and strength might both be expected to increase. They do not actually tend to infinity as $\hat{S}_\theta \to 0$ at peak stress, however. Instead the asymptotic expansion breaks down closer to the peak, and another representation, described in the next section, is required. In fact, as will be seen in Chapter 7, the strength perturbation tends to grow before $\hat{S}_\theta = 0$, but the temperature perturbation actually tends to decrease.

6.2.3 Work Hardening Without Thermal Conductivity: the Response Near Peak Homogeneous Stress

With the same assumptions as above concerning the plastic modulus M and the strain rate sensitivity m, Equation (6.28) for the composite perturbation $\tilde{\lambda} = \tilde{\kappa} - M(t)\tilde{\theta}$ is still valid near the peak homogeneous stress, but only the integral part of Equation (6.29) is needed for temperature because the initial boundary layer can be ignored. With t_m indicating the time at which maximum homogeneous stress occurs, an asymptotic expression for the solution near to t_m may be found by first changing the variable of integration to z, now defined by $e^{-z^2} = \hat{S}(t')/\hat{S}(t_m)$. With this definition there are two branches for $z(t)$ corresponding to times on the two sides of the peak stress. The interval $z(0) \geq z \geq 0$ corresponds to $0 \leq t' \leq t_m$ on the left side of the peak, and $0 \leq z \leq z(t)$ corresponds to $t_m \leq t' \leq t$ on the right side of the peak. Figure 6.3 shows the behavior of the stress response relative to peak stress and the effect of the large power $1/m$. The integral becomes

$$\tilde{\theta}(y, t) \sim -\frac{\tilde{\lambda}_0(y)}{m\, M(0)} \left\{ \int_{z(0)}^{z(t_m)} + \int_{z(t_m)}^{z(t)} e^{-z^2/m} \left[\frac{\hat{S}(t_m)}{\hat{S}(t)} \right]^{1/m} F_\kappa M \frac{dt'}{dz}\, dz \right\}.$$

$$(6.37)$$

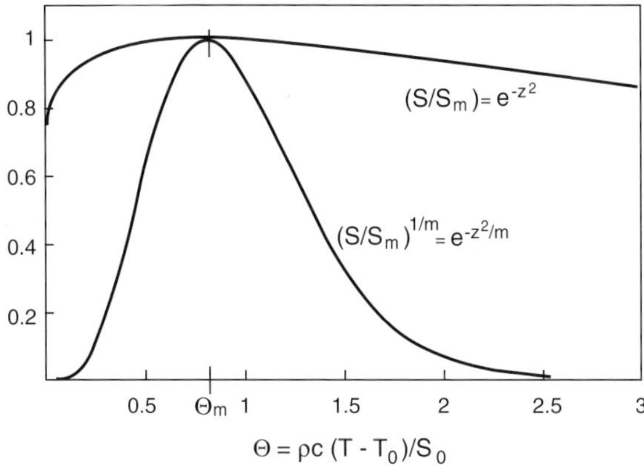

Figure 6.3. The homogeneous response, normalized at peak stress, is shown in the upper curve. The stress ratio shown in brackets in Equation (6.37), when raised to a large power, is a rapidly varying function with vanishing slope at its peak, as shown by the lower curve in the figure. The asymptotic approximation takes advantage of this fact.

The integral has been broken into two parts to take explicit account of the two branches of z. Besides a rapidly varying part as in the previous expansion, the integral has a slowly varying part, which may be expanded into a Taylor series for term by term integration. The first term of the expansion becomes

$$\tilde{\theta}(y,t) \sim -\frac{\tilde{\lambda}_0(y)}{m\,M(0)}\left[\frac{\hat{S}(t_m)}{\hat{S}(t)}\right]^{1/m}(F_\kappa M)_m\left\{\left(\frac{dt'}{dz}\right)_{m^-}\int_{z(0)}^0 e^{-z^2/m}\,dz\right.$$

$$\left.+\left(\frac{dt'}{dz}\right)_{m^\mp}\int_0^z e^{-z^2/m}\,dz\right\}. \tag{6.38}$$

The first integral corresponds to the left side of the peak, and the second integral to the neighborhood of the peak, either to the left $(-)$ or the right $(+)$. Because the lower limit of the first integral is effectively infinite, the value of the integral is $(-1)\sqrt{m\pi}/2$. Similarly, the second integral is $(\sqrt{m\pi}/2)\mathrm{erf}(z/\sqrt{m})$. As before, the value of the derivative dt'/dz may be found by differentiation of the definition for z, but care must be taken to evaluate it on the correct branch of $z(t)$. Differentiating once gives $\hat{S}_\theta(dt'/dz) = -2z$, but because both sides vanish at $z = 0$, it is necessary to differentiate a second time, $\hat{S}_{\theta\theta}\hat{S}(dt'/dz)^2 + \hat{S}_\theta(d^2t'/dz^2) = -2$. Because $\hat{S}_\theta(t_m) = 0$, the second term vanishes at peak stress, and we find that

$$\left(\frac{dt'}{dz}\right)_{m^\mp} = \mp\sqrt{\frac{-2}{(\hat{S}_{\theta\theta}\hat{S})_m}}. \tag{6.39}$$

The minus sign corresponds to the left side of the peak, and the plus sign to the right side.

The curve $\hat{S}(\theta)$ is concave downward at peak stress, so $\hat{S}_{\theta\theta}$ is negative there, and the right hand side of (6.39) is a real number. Equations (6.38) and (6.39) combined give

$$\tilde{\theta}(y,t) \sim -\frac{\tilde{\lambda}_0(y)}{m\,M(0)}\{F_\kappa M\}_m\left\{\frac{-2}{\hat{S}\hat{S}_{\theta\theta}}\right\}_m^{1/2}\frac{\sqrt{m\pi}}{2}\left[1 \mp \mathrm{erf}(z/m^{1/2})\right]e^{z^2/m}, \tag{6.40}$$

where the terms in braces are to be evaluated at peak homogeneous stress, and the minus (plus) sign corresponds to times before (after) peak stress.

Equation (6.35) showed that the perturbation in temperature somewhat before the peak in stress is only $\mathcal{O}(1)$ to leading order in m. Equation (6.40), however, shows that near peak homogeneous stress the perturbation is actually $\mathcal{O}(m^{-1/2})$, which may represent a significant amplification because m is much less than one. These two behaviors are not actually inconsistent with each other because it is a simple matter to show that $\sqrt{m\pi}[1 - \mathrm{erf}(z/\sqrt{m})]/e^{-z^2/m} \to m/z$ as $z \to \infty$ in the right-hand side of (6.40) by L'Hopital's rule. Thus, at some distance before the peak, (6.40) also becomes $\mathcal{O}(1)$ in m. However, after the peak (6.40) shows that the perturbation can be expected to increase rapidly.

Either representation (6.35) or (6.40) may show sufficient growth that linearization of the governing equations is inadequate, but these solutions indicate that the most rapid growth for a sufficiently small perturbation occurs after peak stress. However, a large enough perturbation could conceivably lead to localization before the peak homogenous stress. It is instructive to examine the complete asymptotic solution at peak homogeneous stress according to (6.40).

$$\tilde{\lambda}(y, t_m) = \tilde{\lambda}_0(y)\frac{M(t_m)}{M(0)} + \mathcal{O}(m),$$

$$\tilde{\theta}(y, t_m) \sim -\sqrt{\frac{\pi}{2m}}\frac{\tilde{\lambda}(y, t_m)F_\kappa(t_m)}{\sqrt{-\hat{S}(t_m)\hat{S}_{\theta\theta}(t_m)}} + \mathcal{O}(\sqrt{m}),$$

$$\tilde{\kappa}(y, t_m) \sim \tilde{\lambda}(y, t_m)\left[-\sqrt{\frac{\pi}{2m}}\frac{F_\kappa(t_m)M(t_m)}{\sqrt{-\hat{S}(t_m)\hat{S}_{\theta\theta}(t_m)}} + 1 + \mathcal{O}(\sqrt{m})\right], \tag{6.41}$$

$$\tilde{v}_y(y, t_m) \sim -\frac{\tilde{\lambda}(y, t_m)F_\kappa(t_m)}{m\hat{S}(t_m)} + \mathcal{O}(1).$$

The first of (6.41) reminds us that the composite perturbation does not change sign at lowest order. Thus, points that initially are weaker or warmer than average ($\tilde{\kappa}_0 < 0$ and/or $\tilde{\theta}_0 > 0$), make $\tilde{\lambda}_0 < 0$, and they will be much warmer and more work hardened than average to order $\mathcal{O}(m^{-1/2})$ at peak stress. These points will also have increased strain rate to order $\mathcal{O}(m^{-1})$. Any of these quantities may already be too large to justify linearization of the governing equations, thus suggesting that localization may already have occurred.

6.2.4 Finite Thermal Conductivity and Work Hardening

It is not possible to work out this case in the same detail as the previous two cases. However, it is possible to estimate some of the principal effects of heat conduction. If m is small, the first of (6.18) still indicates an initial boundary layer, at the end of which the first of (6.35) will hold, namely $\tilde{\theta} = -(F_\kappa/\hat{S}_\theta)\tilde{\lambda}$. Now, however, instead of satisfying the second equation of (6.25), after the boundary layer $\tilde{\lambda}$ must be found from a diffusion equation

$$\tilde{\lambda}_t = k\frac{MF_\kappa}{\hat{S}_\theta}\tilde{\lambda}_{yy} + M_\kappa\hat{S}\tilde{\lambda}. \tag{6.42}$$

With the same restrictions on m and M as before, this has the solution

$$\tilde{\lambda}(y, t) = \frac{M(t)}{M(0)} \tilde{\mu}(y, t), \quad \tilde{\mu}(y, 0) = \tilde{\lambda}_0(y), \qquad (6.43)$$

where

$$\tilde{\mu}_\tau = k \tilde{\mu}_{yy},$$

$$\tau = \int_0^t \frac{M F_\kappa}{\hat{S}_\theta} dt'.$$

Not only is the defect parameter compressed by the evolution of M, but diffusion flattens it further on an accelerating time scale because $M F_\kappa / \hat{S}_\theta = 1 + (-F_\theta / \hat{S}_\theta) > 1$ and $\to \infty$ as $\hat{S}_\theta \to 0$. This approximation cannot hold as the peak stress is approached, but it simply illustrates that thermal diffusion tends to suppress the defect.

A more refined analysis, using the Wentzel–Kramers–Brillonin (WKB) method (e.g., see Bender and Orszag 1978) on the Fourier components of (6.18), confirms these results; see Wright (1992). In the region before and after the peak stress, the components may be expressed as a linear combination of two linearly independent solutions,

$$\tilde{\theta}_p \sim \begin{cases} \dfrac{F_\kappa M}{\chi_p} \exp\left\{ -\alpha_p^2 \displaystyle\int^t \dfrac{F_\kappa M}{\chi_p} dt' \right\} \\[2em] e^{-\alpha_p^2 t} S^{-1/m} \exp\left\{ \alpha_p^2 \displaystyle\int^t \dfrac{F_\kappa M}{\chi_p} dt' \right\} \end{cases}, \qquad (6.44)$$

where for $p = 1, 2, \ldots$, we have $\alpha_p = \sqrt{k} p\pi$ and $\chi_p(t) = \hat{S}_\theta(t) + m\alpha_p^2 + m(d/dt)(\ln F_\kappa M)$. The first of (6.44) corresponds to (6.43), and the second corresponds to the initial boundary layer. After the homogeneous stress passes the maximum and begins to decrease, $\chi_p(t)$ may pass through a zero and become negative. As in the previous section, a more refined analysis is required near these zeros. It may be shown that the Fourier components in the temperature solution behave essentially as Whittaker functions near the zero in $\chi_p(t)$. After the change of sign, the lowest Fourier components begin to grow exponentially in sequence, beginning with the first at $p = 1$. The effect of heat conduction is to retard the effective peak for each Fourier component. Furthermore, although the slope of the homogeneous response becomes negative, it remains finite, and therefore, because α_p^2 grows as the square of the integer p, only a finite number of Fourier components will begin to grow.

All of the three cases above have been worked out for $m = \text{const}$ and $M_\theta = 0$, but the solutions can be generalized to remove both restrictions, as was partially done in Wright (1992). The general character of the response does not change.

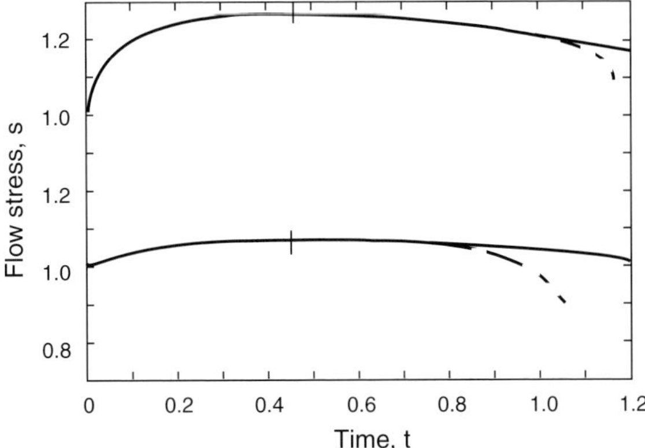

Figure 6.4. Homogeneous response curves for two materials, one with strong work hardening and one with weak work hardening. Peak stresses are indicated by the tick marks. The dashed curves show the start of stress collapse caused by a typical, small perturbation. (Reproduced from Wright 1992, with permission from Elsevier Science.)

6.2.5 Discussion of Quasi-Static Solutions

Figures 6.4, 6.5, 6.6, 6.7, and 6.8 illustrate the presence of boundary layers in the solution to the linearized perturbation equations with both work hardening and heat conduction included. Figure 6.4 shows the evolution of the adiabatic stress in two cases; the lower one with weak work hardening and the upper

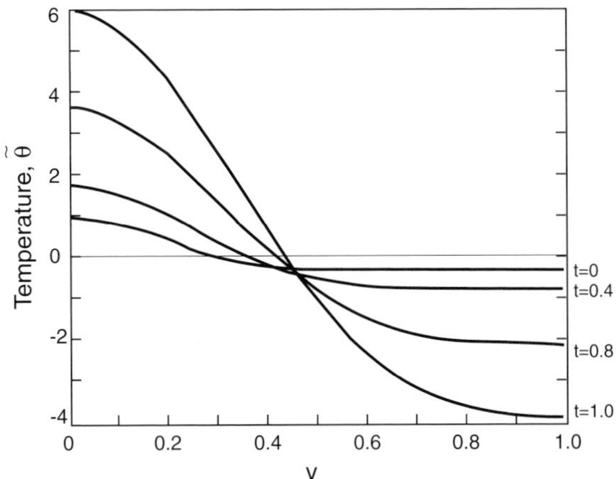

Figure 6.5. Cross sections of an evolving temperature perturbation according to the linearized equations for the weakly work hardening case in Figure 6.4. The initial amplitude at $y = 0$ was normalized to have the value $\tilde{\theta}(0, 0) = 1$. The accelerating growth of the profile relative to the homogeneous growth in temperature is clearly visible. (Reproduced from Wright 1992, with permission from Elsevier Science.)

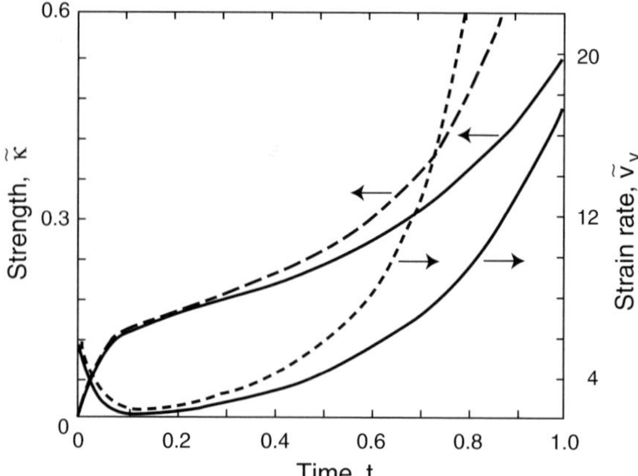

Figure 6.6. Growth of perturbations in strength and strain rate for the weakly work hardening material in the center of a developing shear band. The initial perturbation in strength was zero; the initial perturbation in temperature was the same as in Figure 6.5. The solid lines correspond to the linearized equations, and the dashed lines correspond to the nonlinear equations. Note the initial boundary layers in time, as predicted asymptotically in Equations (6.31)–(6.33). Peak stress occurs at approximately 0.44. (Reproduced from Wright 1992, with permission from Elsevier Science.)

one with strong work hardening. The homogeneous response after peak stress is similar in both cases, showing a slow decrease. The nonlinear response of stress to a small imperfection is also shown in both cases. At some time after the peak the stress begins to decrease rapidly. Note that the departure from the

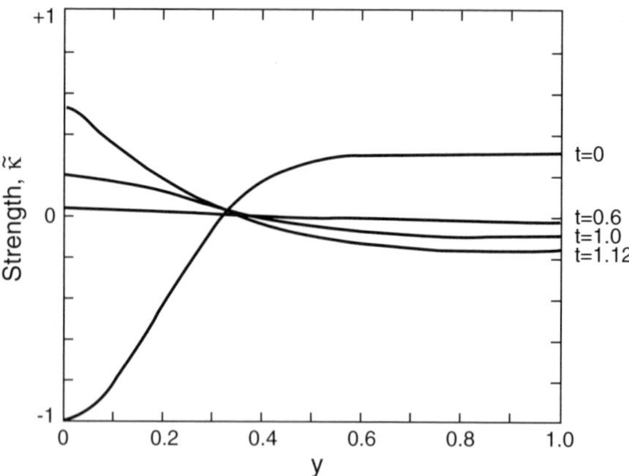

Figure 6.7. Cross sections of an evolving strength perturbation according to the linearized equations for the strongly work hardening case in Figure 6.4. The initially weak region is becoming a region of strength, even though the stress has not fully collapsed nor full localization occurred at $t = 1.12$, as may be seen by referring to the top curve in Figure 6.4. (Reproduced from Wright 1992, with permission from Elsevier Science.)

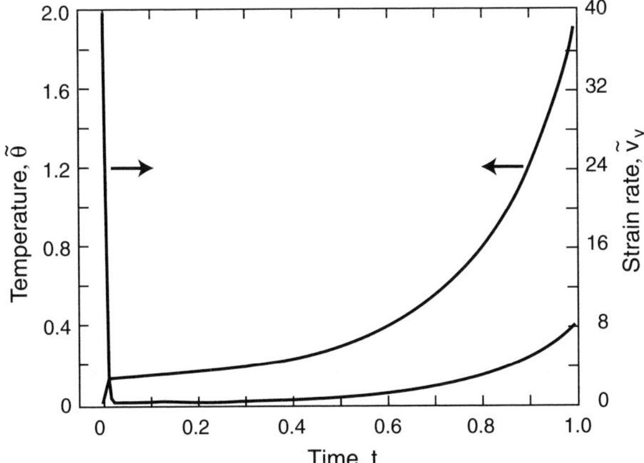

Figure 6.8. Evolution of temperature and strain rate for the strongly work hardening material in the center of a developing shear band. The initial boundary layer is much more rapid in this case than in the case of the weakly work hardening material. (Reproduced from Wright 1992, with permission from Elsevier Science.)

homogeneous curve begins well after peak stress, indicated by the tick mark, in both cases.

For the case of weak work hardening, Figure 6.5 shows the evolving profile of a pure temperature perturbation at several times after initiation. Figure 6.6 shows how the work hardening and strain rate in the center of the perturbation respond for the same case. Note the rapid initial boundary layer in time, as predicted, followed by accelerating growth. The lower curve in Figure 6.4 shows that peak stress occurs at the tick mark at approximately $t = 0.44$. The growth shown in 6.6, however, begins well before the peak, continues smoothly through the peak, and continues to accelerate. For comparison the dashed lines show the true nonlinear response in Figure 6.6 also. Note how the linearized calculation underestimates the true response, particularly for the strain rate, although its overall character is much the same as the nonlinear response. There does not seem to be any particular time when the linearized response fails, but by the time that significant stress departures actually begin near $t = 0.8$, the discrepancies are large.

For the case of strong work hardening, Figure 6.7 shows the evolving profile of a pure strength perturbation at several times after initiation. Note how the initial region of weakness has reversed sign, as predicted, and become a region of higher strength by the end of the calculation. Figure 6.8 shows how the temperature and strain rate in the center of the perturbation evolve for this case. Note the extremely rapid boundary layer. Again there is slow, but accelerating, growth of the perturbation as soon as the boundary layer is past, and growth continues smoothly through the time of peak stress as in the previous case. Although Figure 6.4 shows that the stress departs from the homogeneous response near $t = 1.0$, according to the nonlinear prediction, the linearized responses for temperature and strain rate show significant growth well before that time. Again, however, there does not seem to be any particular time when the linearized response fails.

6.2.6 Scaling Laws and Scaling Parameters

The perturbed flow law in Equation (6.10) can be used to suggest a possible form for a scaling law that relates a scaled defect to a scaled strain or temperature at failure. The basic idea, suggested by H. Ockendon (1990, personal communication), follows from the observation that when the maximum value of the perturbed strain rate becomes comparable to the imposed, homogeneous strain rate, then the linearized equations can no longer be justified as a reasonable approximation to the fully nonlinear system. That is to say, linearization has surely failed when $\tilde{v}_y = 1$, or from (6.10), when t is such that

$$\tilde{v}_y = -\frac{F_\kappa}{F_{\dot\gamma}}\tilde{\kappa} - \frac{F_\theta}{F_{\dot\gamma}}\tilde{\theta} = 1. \tag{6.45}$$

This criterion does not determine a "time of failure" in any strict sense, but it does give an order-of-magnitude estimate, and it does tend to isolate key scaling factors.

For example, suppose that the flow law for a perfectly plastic material (in nondimemsional form) with linear thermal softening is given by $F = (1 - a\theta)v_y^m$. With $v_y = 1$ on the homogeneous solution, we have $F_\kappa = 0$, $F_\theta = -a$, $F_{\dot\gamma} = mF$, $\hat{S}(\Theta) = 1 - a\Theta = \exp(-at)$, and $\exp\{-\int_0^t (\hat{S}_\theta/m)\,dt'\} = e^{at/m}$. As a result, (6.45) with (6.21) and $y = 0$ yields

$$\frac{a}{m}\frac{\tilde{\theta}(0,t)}{1 - a\Theta(t)} = \frac{a}{m}\frac{e^{at/m}}{e^{-at}}\sum_1^\infty \theta_n e^{-k(n\pi)^2 t} = 1. \tag{6.46}$$

With use of only the first term of the Fourier series, Equation (6.46) identifies an approximate "critical time" as

$$t_{\rm cr} \approx \frac{1}{[(1 + m)/m]a - k\pi^2}\ln\frac{m}{a\theta_1} \tag{6.47}$$

as first noted by H. Ockendon (1990, personal communication). When expressed in dimensional terms, after it is recognized that $t_{cr} = \gamma_{cr}$, (6.47) becomes

$$\gamma_{cr} \approx \left(\frac{\bar{a}s_0}{m\bar{\rho}c}\right)^{-1}\left(1 + m - \frac{\bar{k}\pi^2}{\bar{a}s_0\dot{\gamma}_0 H^2}\right)^{-1}\ln\frac{m}{\bar{a}\bar{\theta}_1}. \tag{6.48}$$

The ratio $a/m = \bar{a}s_0/m\bar{\rho}c$, where $s_0 = \kappa_0(b\dot{\gamma}_0)^m$, might be called the "susceptibility" to adiabatic shear for a perfectly plastic material, because the critical strain is inversely proportional to that ratio. This basic susceptibility is modified by heat conduction (relative to the rate of thermal softening), which will delay the critical strain at lower strain rates, as was observed in the parametric calculations shown in Figure 5.5. At high strain rates, however, the effect of heat conduction becomes very small. An "imperfection sensitivity" may be associated with the coefficient of θ_1 in (6.47). For a perfectly plastic material this is $\bar{a}/m$ in dimensional terms.

Similarly, Equation (6.45), rewritten in the form

$$-\frac{\hat{S}_\theta}{F_{\dot{\gamma}}}\tilde{\theta} - \frac{F_\kappa}{F_{\dot{\gamma}}}\tilde{\lambda} = 1, \tag{6.49}$$

may also be used to find scaling factors for a work hardening material if heat conduction is ignored. Near the peak homogeneous stress, $\hat{S}_\theta = (\hat{S}_{\theta\theta})_m(\theta - \theta_m)$, so that with (6.28) and (6.40) Equation (6.49) becomes

$$-\frac{\{\hat{S}_{\theta\theta}\}(\theta - \theta_m)}{m\hat{S}}\left\{-\frac{\tilde{\lambda}F_\kappa}{m}\right\}\left\{\frac{-2}{\hat{S}\hat{S}_{\theta\theta}}\right\}^{1/2}\frac{\sqrt{m\pi}}{2}\left[1 \mp \text{erf}(z/m^{1/2})\right]e^{z^2/m} - \frac{F_\kappa\tilde{\lambda}}{m\hat{S}} = 1. \tag{6.50}$$

As before, terms in braces are to be evaluated at peak homogeneous stress. This equation appears to be too complicated for easy interpretation, but it can be simplified considerably by recognizing key scaled variables. To that end, let a measure of the defect be given by

$$\beta = -\{F_\kappa\tilde{\lambda}/m\hat{S}\}, \tag{6.51}$$

define the distance from peak homogeneous stress by $\eta = z/\sqrt{m}$, and use the exact definitions $m\hat{S} = F_{\dot{\gamma}}$ and $\eta^2 = z^2/m = -\int_{t_m}^{t}(\hat{S}_\theta\hat{S}/F_{\dot{\gamma}})\,dt$, which may be approximated by

$$\eta^2 \approx -\frac{1}{2}\left\{\frac{\hat{S}_{\theta\theta}}{F_{\dot{\gamma}}}\right\}(\Theta - \Theta_m)^2 \approx -\frac{1}{2}\left\{\frac{\hat{S}^2\hat{S}_{\theta\theta}}{F_{\dot{\gamma}}}\right\}(\gamma - \gamma_m)^2$$

$$= -\frac{1}{2}\left\{\frac{\hat{S}\hat{S}_{\theta\theta}}{m}\right\}(\gamma - \gamma_m)^2. \tag{6.52}$$

Then for times past the peak homogeneous stress equation (6.50) may be written

$$\eta e^{\eta^2}(1 + \operatorname{erf}\eta) = (1 - \beta)/(\sqrt{\pi}\beta) \tag{6.53}$$

Because of the nature of the approximations made in obtaining (6.53), it should be most accurate near peak homogeneous stress, or near $\eta = 0$. In any event it suggests that localization occurs at a scaled time (measured in terms of homogeneous temperature or strain) that depends only on the magnitude of a scaled composite defect. That is to say,

$$\eta = f(\beta). \tag{6.54}$$

Note that the definition for the scaled strength of a defect as given by (6.51), is the same as the nondimensional perturbation of strain rate, as given by the fourth equation of (6.41).

Given the complexity of the full nonlinear problem, Equation (6.54) makes the remarkable claim that if appropriate, physically determined scale factors are used, then localization will occur at an increment of nominal strain past peak homogeneous stress that depends only on the largest negative value of the initial composite defect, $\tilde{\lambda}_0 = \tilde{\kappa}_0 - M(0)\tilde{\theta}_0$. The point of localization may also be expressed as an increment of time or of temperature. Furthermore, if the scaled defect is greater than one, $\beta > 1$, then localization may actually occur before peak homogeneous stress.

These predicted features may easily be tested numerically, as was done by Wright (1994) for one simple material model with power law work hardening, exponential thermal softening, and power law rate hardening. Twelve cases were run with a wide variety of physical and applied properties, but when plotted with appropriate scaling factors, the results clustered just below the curve

$$\eta = \pi^{-1/2}\ln\beta^{-1} \tag{6.55}$$

rather than along the curve (6.53) itself, as shown in Figure 6.9. It is easily worked out that the two curves given by (6.53) and (6.55) are tangent to each other at the peak of the homogeneous stress where $\beta = 1$. Thus, as expected, (6.53) appears to be a good approximation only near the peak, but the scaling deduced from the linearized theory actually works quite well. It is also amply confirmed that the initial composite defect, rather than the individual defects, is all that is required up to ~ 0.035; see Wright (1994).

In dimensional terms Equation (6.55) reads

$$\gamma_{cr} - \gamma_m = \sqrt{\frac{2mn}{(1+n)\pi}\frac{\bar{\rho}c}{\bar{a}S_m}}\,\ln\frac{m/\bar{a}}{\bar{\theta}_0 - (\gamma_0/n\bar{\rho}c)\bar{\kappa}_0} \tag{6.56}$$

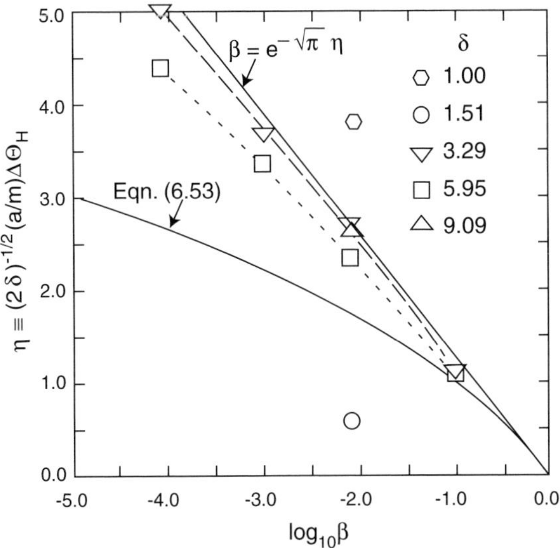

Figure 6.9. Linearized equations suggest that the relation between properly scaled defect and collapse temperature (homogeneous temperature past peak homogeneous stress where the stress collapses) is given by the lower curve. Numerical experiments with the fully nonlinear equations show that collapse occurs shortly before the upper curve, which is tangent to the lower curve at the origin. The parameter $\delta = n/m(1+n)$ has only a weak influence on the results. (Reproduced from Wright 1994, with permission from Elsevier Science.)

Now compare Equations (6.56) and (6.48) without thermal conductivity. The two have strong similarities: the ratio of strain rate sensitivity to thermal sensitivity plays a strong role in both cases, but in the case of work hardening the ratio of strain rate sensitivity to the work hardening exponent modifies the logarithmic prefactor.

A rough estimate of the susceptibility may be made by setting the logarithmic term to one (in effect choosing the scaled defect to be moderately strong; $\beta = e^{-1}$) and then approximating the leading multiplier by $(m/a)\sqrt{n/m}$. Then because the strain at peak stress often has the approximate value n/a (see Wright 1992), the critical strain in the work hardening case may be estimated as $\gamma_{cr} \approx (n/a) + (m/a)\sqrt{n/m}$, as compared with $\gamma_{cr} = m/a$ for the case of perfect plasticity. Combining the two estimates leads to $\gamma_{cr} \approx (m/a)\max\{1, (n/m + \sqrt{n/m})\}$. Let the susceptibility be defined as the reciprocal of this estimate for critical strain so that

$$\frac{\chi_{SB}}{a/m} = \min\left\{1, \frac{1}{(n/m) + \sqrt{n/m}}\right\}. \tag{6.57}$$

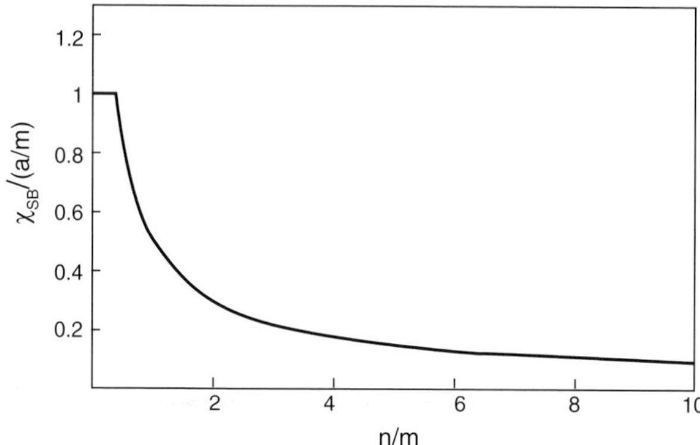

Figure 6.10. Susceptibility to shear band formation, χ_{SB}, as estimated by Equation (6.57), is proportional to the ratio of nondimensional thermal sensitivity to strain rate sensitivity times a decreasing function of the ratio of the work hardening exponent to the strain rate sensitivity. In dimensional terms, $\chi_{SB} = (\frac{1}{\rho c}\frac{\partial s}{\partial \theta} / \frac{\partial \ln s}{\partial \ln \dot{\gamma}}) f(n/m)$, where $f(\cdot)$ is the function shown in the figure.

Equation (6.57) was first given by Wright (1992) and is plotted in Figure 6.10. At low values of the parameter n/m the susceptibility is essentially the same as for a perfectly plastic material, but for materials with work hardening that is relatively stronger than rate hardening, the susceptibility becomes lower. There should also be a transitional region between the two extreme behaviors, but the shape is unknown. In any event the figure must be regarded as qualitative only. More refined estimates could be made, but the figure captures the essential ideas. Analysis indicates that the main factor in determining susceptibility is the ratio of thermal sensitivity to strain rate sensitivity, but in the case of work hardening that basic estimate must be modified by a function of a single parameter.

In the literature of metal forming (e.g., see Semiatin and Jonas 1984), an alpha parameter has been defined as the ratio of the work softening slope to the strain rate sensitivity,

$$\alpha = \frac{|d\bar{s}/d\varepsilon|}{m\bar{s}}, \tag{6.58}$$

where generalized flow stress and strain are used. But because $\bar{s}d\varepsilon = \rho c\,dT$ in the adiabatic case, and because $a \equiv -\bar{s}_T/\rho c$, their alpha parameter is the same as the ratio a/m as used in this book, although there are some nuances in the way they actually make use of their parameter.

Molinari and Clifton (1987) also predicted a logarithmic dependence on defect size for the strain at localization. Duffy and Chi (1992) carefully prepared

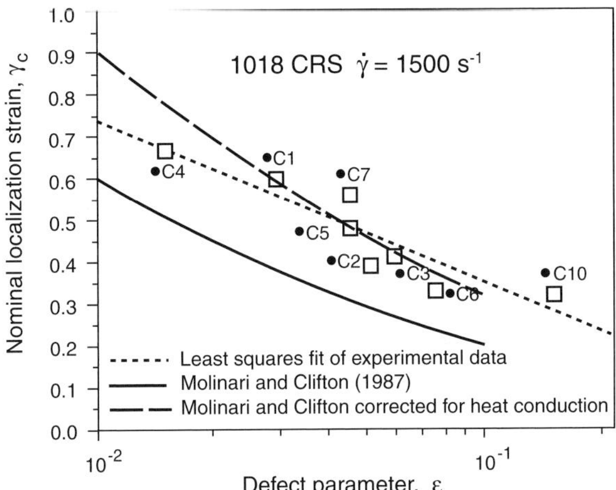

Figure 6.11. The localization strain decreases as the defect increases. The lower curve and the data points are reproduced from Duffy and Chi (1992, with permission from Elsevier Science). The dashed curve is a correction to the theoretical curve for finite thermal conductivity, as suggested by Equation (6.48).

torsion specimens with known defects of wall thickness and measured the strain at localization, as shown in Figure 6.11. Their results followed the Molinari and Clifton prediction, but with an additional delay beyond the theoretical result. Duffy and Chi made no allowance for the effect of heat conduction, but if the indicated theoretical curve is multiplied by the additional factor indicated in Equation (6.48), $[1 + m - (\bar{k}\pi^2/\bar{a}s_0\dot{\gamma}_0 H^2)]^{-1}$, then with reasonable values estimated for the physical constants $\bar{k} = 48$ W/mK, $\bar{a} = 1/900$ K^{-1}, $s_0 = 0.5$ GPa, $\dot{\gamma}_0 = 1500\,\mathrm{s}^{-1}$, $H = 1.25$ mm, and $m = 0.03$, the correction factor is approximately 3/2, and the data lie squarely on top of the corrected theoretical curve. The use here of the correction factor is somewhat ad hoc in character because the term was not derived for a work hardening material.

6.3 Infinite Domain: Band Spacing and Patterning

Figure 5.5 showed the variation of critical strain with increasing nominal strain rate in a rigid, perfectly plastic material. The curve has a U shape, tending to infinity either at a low or infinite strain rate and having a single minimum in the middle. Numerical experiments have shown that a similar U curve also occurs for work hardening materials (Klepaczko, Lipinski, and Molinari 1988; also unpublished work of Walter 1987, as shown in Wright 1994). At low strain rates heat conduction retards localization, and at high rates inertia retards localization. Perturbation analysis can be used to estimate the location of the minimum in a finite region, and a slight variation of the analysis can also be used to estimate the most likely minimum spacing for nucleation of shear bands in an infinite domain.

Following Wright and Ockendon (1996), in the case of a material that is assumed to soften linearly with temperature and to harden with strain rate according to a power law, that is $s = (1 - a\theta)v_y^m$, the homogeneous solution is $\hat{S} = 1 - a\Theta = e^{-at}$ and $V = y$ in nondimensional form, so that the perturbation equations, including inertia, become

$$\tilde{s}_y = \rho\tilde{v}_t,$$

$$\tilde{\theta}_t = k\tilde{\theta}_{yy} + \tilde{s} + e^{-at}\tilde{v}_y, \tag{6.59}$$

$$\tilde{s} = me^{-at}\tilde{v}_y - a\tilde{\theta}.$$

The last two equations may be solved for the perturbations of stress and strain rate in terms of the perturbation in temperature:

$$\tilde{s} = \frac{m}{1+m}\left[\tilde{\theta}_t - k\tilde{\theta}_{yy} - \frac{a}{m}\tilde{\theta}\right],$$

$$\tilde{v}_y = \frac{e^{at}}{1+m}\left[\tilde{\theta}_t - k\tilde{\theta}_{yy} + a\tilde{\theta}\right]. \tag{6.60}$$

When the first of (6.59) is differentiated by y and then (6.60) is used, the balance of linear momentum becomes

$$m\left[\tilde{\theta}_t - k\tilde{\theta}_{yy} - \frac{a}{m}\tilde{\theta}\right]_{yy} = \rho\left[e^{at}(\tilde{\theta}_t - k\tilde{\theta}_{yy} + a\tilde{\theta})\right]_t. \tag{6.61}$$

Once again the perturbation equation has a time-dependent coefficient, so the method of frozen coefficients is not necessarily reliable. In this particular case, a Fourier decomposition in the spatial coordinate may be assumed for $\tilde{\theta}$, which will have time-dependent Fourier components:

$$\tilde{\theta}(y, t) = \sum_{n=1}^{\infty} F_n(t)\cos(n\pi y). \tag{6.62}$$

To satisfy (6.61) the Fourier components in (6.62) must satisfy

$$-m(n\pi)^2[\dot{F}_n + k(n\pi)^2 F_n - (a/m)F_n] = \rho\frac{d}{dt}\{e^{at}[\dot{F}_n + k(n\pi)^2 F_n + aF_n]\}. \tag{6.63}$$

Equation (6.63) may be put in simpler form by letting $F_n = u_n w_n$ where u_n is chosen so as to eliminate $\dot{u}_n$ in the new equation for w_n. After some manipulation we have

$$u_n = \exp\left\{ -at - \frac{1}{2}k(n\pi)^2 t - \frac{1}{2}\frac{m}{\rho a}(n\pi)^2(1 - e^{-at}) \right\},$$

$$\ddot{w}_n - \left(\frac{n^2\pi^2}{2\rho}\right)^2 \left[(k\rho - me^{-at})^2 + \frac{4a\rho}{n^2\pi^2}\left(1 + \frac{m}{2}\right)e^{-at} \right] w_n = 0.$$

(6.64)

In this form the second equation of (6.64) is in standard form for WKB analysis (e.g., see Bender and Orszag 1978) with $n^2\pi^2/2\rho$ taken to be a large parameter.

The standard WKB solution that grows with time may be written as

$$w_n = A_n^{-1/4} \exp\left\{ \frac{n^2\pi^2}{2\rho} \int A_n^{1/2}\, dt \right\},$$

$$A_n = (me^{-at} - k\rho)^2 + \frac{4a\rho}{n^2\pi^2}\left(1 + \frac{m}{2}\right)e^{-at}.$$

(6.65)

The integral in (6.65) may be evaluated exactly by letting $x = e^{-at}$ and using tables of integrals, but the result is too complicated for easy interpretation. At early times, when the linearized equations are expected to be most accurate, a good representation for the nth Fourier component of strain rate, as computed from the second equation of (6.60), after some labor may be reduced to the deceptively simple expression

$$\dot{\gamma}_n \approx \varepsilon_n \frac{a}{m} e^{\alpha_n t},$$

(6.66)

where the coefficient in the exponential is given by

$$\alpha_n = n^2\pi^2 \frac{m}{2\rho} \left\{ \left[\left(1 - \frac{k\rho}{m}\right)^2 + \frac{4 + 2m}{n^2\pi^2}\frac{a\rho}{m^2} \right]^{1/2} - 1 - \frac{k\rho}{m} + \frac{1}{n^2\pi^2}\frac{a\rho}{m} \right\}$$

(6.67)

and the initial value of the nth Fourier component of temperature is ε_n. Although (6.66) has the same exponential form as that *assumed* in the method of frozen coefficients, here it has been *derived* as an approximation that is valid for early times. The last term in (6.67) comes from the $A_n^{-1/4}$ term in the WKB approximation.

Various scaling laws now follow by maximizing (6.67) with respect to the number of the Fourier component, n, the characteristic length scale, H, or the applied strain rate, $\dot{\gamma}_0$. The number of the Fourier component is stated explicitly in (6.67), but the length scale is embedded in the parameters (ρ, k), although the product ρk is independent of H. The applied strain rate is embedded in the parameters (ρ, k, a), as may be seen by referring back to the definitions in (6.4) and (6.6), and after recalling that $s_0 = \bar{F}(\kappa_0, 0, \dot{\gamma}_0)$. For example, in the present case, we have $s_0 = \kappa_0(b\dot{\gamma}_0)^m$.

In designing an experiment it would be useful to be able to estimate the maximum number of shear bands that are likely to appear or, as in a torsional Kolsky bar experiment, to design the length scale (gage length) so that only one band can appear. Conversely, if the length scale is large so that it is possible to nucleate a large number of bands, it would be useful to be able to estimate the most likely minimum spacing for nucleation of shear bands. These two questions may be examined together because the Fourier number and the length scale always appear in the combination $H/n\pi$ through the term $\rho/n^2\pi^2$.

If the length scale is fixed (the gage section in a test specimen, for example), then the maximum number of bands to be expected is that which will maximize the growth rate, α_n, in (6.67). Conversely, with $n = 1$ identifying the longest wavelength, maximization of α_n in (6.67) with respect to H will select the half-wavelength that grows most rapidly. The basic idea to be exploited here is that shear bands do not propagate laterally but grow in the material on a fixed material plane. Actually the minimum spacing for nucleation and the minimum for longer-term survival may be separate questions. If multiple perturbations nucleate too closely together, a dominant one can absorb the weaker ones into its own structure before a fully formed band can develop, as shown computationally by Bai (1989), but if bands are nucleated far enough apart, they can continue to grow independently more or less at the material location where they were first nucleated, as shown by Kwon and Batra (1988). The linearized analysis developed in this chapter can only follow the beginning stages of growth following the initial nucleation; however, it still has the capability to identify the earliest, gross features of patterning. Therefore, it may be argued that two independently nucleating bands have the best chance of remaining independent if they are more widely separated than the wavelengths that grow most rapidly in the initial stage.

The analysis is simplified by adopting a different parameterization. Thus, we let

$$\xi \equiv \frac{k\rho}{m} = \frac{\bar{k}\dot{\gamma}_0}{mcs_0}, \quad \eta \equiv \frac{1}{2}\frac{(2+m)}{n^2\pi^2}\frac{a\rho}{m^2} = \frac{1}{2}\frac{2+m}{m^2}\frac{\bar{a}\dot{\gamma}_0^2}{c}\left(\frac{H}{n\pi}\right)^2 \tag{6.68}$$

so that the ratio $H/n\pi$ is isolated in the term, η. Equation (6.67) now becomes

$$\alpha_n = \frac{2+m}{4}\frac{a}{m}\left\{\frac{[(1-\xi)^2 + 4\eta]^{1/2}}{\eta} - \frac{1+\xi}{\eta}\right\} + \frac{a}{2}. \tag{6.69}$$

Setting $d\alpha_n/d\eta = 0$ leads to the algebraic condition $\eta^2 - 4\xi\eta - \xi(1-\xi)^2 = 0$, which has the exact solution $\eta = \xi^{1/2}(1 + \xi^{1/2})^2$. The same relation occurs for maximization on either n or H.

For example, when interpreted as the largest number of bands expected to nucleate in the gage length $L = 2H$, the result, when second-order terms are ignored, is

$$n = \frac{L}{2\pi}\left[\frac{\dot{\gamma}_0^3 \bar{a}^2 s_0}{m^3 \bar{k} c}\right]^{1/4}. \tag{6.70}$$

In general (6.70) will not yield a whole number, of course, but it should tell the experimenter approximately what to expect from a particular experimental design. For example, if $\dot{\gamma}_0 = 10^3 \, \text{s}^{-1}$, $\bar{a} = 8 \times 10^{-4} \, °\text{C}^{-1}$, $s_0 = 0.8 \, \text{GPa}$, $m = 0.02$, $\bar{k} = 50 \, \text{W/mC}$, and $c = 500 \, \text{J/kg} \, °\text{C}$ (these are approximate numbers for a strong steel), then with a gage length of $L = 1$mm, the result is only $n = 0.2$. This tells the experimental designer that it may be difficult to form a shear band unless the gage length or the nominal strain rate is increased. Certainly the applied strain rate is larger than the minimum required to form one band. According to the discussion following (6.23), the cutoff, below which no bands can form, occurs when $a/m = k\pi^2$. In dimensional terms the cutoff rate is $\dot{\gamma}_0 = 4\pi^2 m\bar{k}/\bar{a}L^2 s_0$ or, with the numbers used above, $\dot{\gamma}_0 \approx 62 \, \text{s}^{-1}$. Probably the design point of $1000 \, \text{s}^{-1}$ is to the left of the minimum in the U curve where heat conduction can strongly delay the formation of the shear band. This point will be examined further below.

In a larger specimen where many bands may form, the maximum in (6.69) may be interpreted as giving the least spacing to be expected between nucleating shear bands. The result, given by Wright and Ockendon (1996), is

$$L = 2\pi\left[\frac{m^3 \bar{k} c}{\dot{\gamma}_0^3 \bar{a}^2 s_0}\right]^{1/4}. \tag{6.71}$$

With the same numbers as above the minimum spacing for nucleating shear bands is estimated to be approximately 5.0 mm. At a nominal strain rate of $30,000 \, \text{s}^{-1}$, the estimated spacing becomes approximately 0.4 mm.

Other authors have taken different approaches toward estimating the expected spacing between shear bands. Grady and Kipp (1987) argued that localization in a shear band causes unloading to diffuse away from the site and to quench localization in other nearby sites. By comparing the time to localize with the time for diffusive unloading in a rate-independent material, they estimated that to avoid mutual interference of this kind, bands must be separated by at least

$$L = 2\left[\frac{9\bar{k} c}{\dot{\gamma}_0^3 \bar{a}^2 \kappa_0}\right]^{1/4}, \tag{6.72}$$

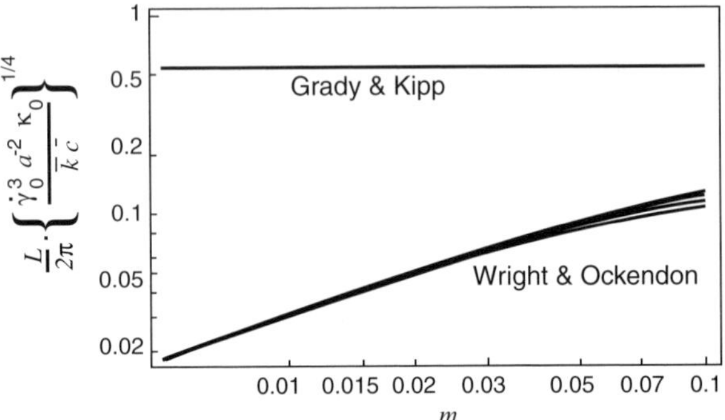

Figure 6.12. Comparison of the minimum spacing between shear bands as predicted by Grady and Kipp (1987) and by Wright and Ockendon (1996). (Reproduced from Wright and Ockendon 1996, with permission from Elsevier Science.)

where κ_0 is a static strength of the material. Except for numerical factors, the strain rate sensitivity, and the use of static strength versus the dynamic strength, these two estimates are very similar. Figure 6.12 compares (6.71) and (6.72). The estimate from (6.72) is typically five to ten times that from (6.71).

Still another approach has been taken by Molinari (1997). Proceeding by linearization about a homogeneous solution and by the method of frozen coefficients, he argues further that not only should the maximum rate of growth at the initial instant be considered, but also that the absolute maximum growth rate for all times on the homogeneous solution should be found. For the case of linear softening in a perfectly plastic material, he arrives at

$$L = 2\pi \left[\frac{m^3 \bar{k} c (1 - \bar{a}\bar{\theta})}{\dot{\gamma}_0^3 \bar{a}^2 s_0} \right]^{1/4} , \tag{6.73}$$

and again the scaling is similar except for the added temperature factor, which makes this estimate less than (6.71). Molinari (1997) also applied his method to work hardening materials, but because of complexity, only numerical results could be given.

Although the method of frozen coefficients may work for the simple case of linear thermal softening in the absence of work hardening, as suggested by the approximate expression (6.66), it is hard to understand how it could be effective when the linearized equations have boundary layers, as in (6.31), (6.32), and (6.33). There may be some merit in considering growth rates at times beyond the initial time, however, because weak perturbations may not produce actual localization until substantial nominal strain has occurred with its corresponding plastic heating and thermal softening of the material. However, none of these estimates account for the strength of the perturbation at all, so perhaps it is just as well to use only the initial rate of growth. In any event, the most important point on which all estimates agree is that an important length scale is proportional to $(\bar{k}c/\dot{\gamma}_0^3\bar{a}^2 s_0)^{1/4}$ where s_0 is some estimate of the strength, either static or dynamic.

Wright and Ockendon (1996) also used Equation (6.67) to estimate the minimum of the U curve, as in Figure 5.5, with respect to the nominal strain rate, $\dot{\gamma}_0$. The argument is essentially the same as above, namely that the minimum should occur at the strain rate that makes the fundamental wavelength grow the fastest. That is, find $\dot{\gamma}_0$ that makes $d\alpha_1/d\dot{\gamma}_0 = 0$. In practice the algebra is much more involved than that given previously, because strain rate enters into the expression in more complex ways than does the length scale. In addition, in finding the terms of dominant balance in the resulting algebraic expression, it is necessary to examine three terms. The final expression is a cubic, which may be solved by using Cartan's rule. After neglecting small quantities, the result in dimensional terms is

$$\dot{\gamma}_0 = \frac{m}{2^{1/3}}\left[\left(\frac{2\pi}{L}\right)^4 \frac{\bar{k}c}{\bar{a}^2 s_0}\right]^{1/3} + \frac{m^2}{3\times 2^{2/3}}\left[\left(\frac{2\pi}{L}\right)^2 \frac{c^2 s_0}{\bar{k}\bar{a}}\right]^{1/3}. \tag{6.74}$$

In using (6.74) one should remember that the dynamic strength is given by $s_0 = \kappa_0(b\dot{\gamma}_0)^m$, and therefore it contains the unknown nominal strain rate. However, because the strain rate sensitivity is small, a single iteration will produce a very good solution. When applied to the case shown in Figure 5.5, the estimate given for the strain rate that will produce the minimum critical strain is approximately $2500\ \text{s}^{-1}$. This appears to be a very accurate estimate.

When the same properties as those following (6.70) are used, the strain rate that maximizes the growth rate and hence minimizes the strain at extreme localization is $\dot{\gamma}_0 \approx 11600\ \text{s}^{-1}$. Thus, the previously assumed strain rate $\dot{\gamma}_0 = 1000\ \text{s}^{-1}$ that corresponded to $n = 0.2$ does indeed lie to the left of the minimizing strain rate, as previously conjectured. In any case, with these calculations the experimentalist would be forewarned to expect that heat conduction will strongly influence the experimental results.

As a second example, suppose $\bar{a} = 1500^{-1}\ \text{C}^{-1}$, $s_0 = 0.4\ \text{GPa}$, $m = 0.05$, $\bar{k} = 20\ \text{W/mC}$, and $c = 520\ \text{J/kg C}$. These are approximate numbers for Ti, as found in tables and response data given by Bai and Dodd (1992). Now the strain rate that produces the minimum strain to localization for a half-gage length of $L = 1.25$ mm is approximately $\dot{\gamma}_0 = 44.2 \times 10^3\ \text{s}^{-1}$, according to (6.74). If the design strain rate is $1500\ \text{s}^{-1}$, then the expected number of Fourier components is still $n \approx 0.2$ as before. Now, however, it should be expected that heat conduction has a much greater influence than in the previous example, even though the thermal conductivity is less than half the previous value because the design point is so far to the left of the minimum of the U curve.

As a final remark, equation (6.70) indicates that increasing the gage length for a fixed strain rate will increase the largest number of bands to be expected toward 1.0. At the same time, equation (6.74) indicates that increased gage length will decrease the strain rate at which most rapid localization occurs. To reduce the influence of heat conduction on an experiment, the applied strain rate should be as close to the rate for most rapid localization as possible. However, as the gage length becomes larger, it becomes more difficult to achieve equilibrium in the experiment.

6.4 Concluding Remarks

Because many of the results in this chapter depend on a reasonably careful mathematical argument, it is worthwhile to review the approach and to restate some of the principal results just in words. The intent has been to examine the material stability of viscoplastic solids in high rate shearing in which strain softening has set in as a result of the dominance of heating from plastic working. Actually a more restricted problem has been examined, namely the stability of homogeneous shearing, rather than stability of a general motion. A standard mathematical approach was taken, namely to examine the behavior of small perturbations in a system of equations that have been linearized about a homogeneous solution to the full nonlinear set of equations. These linearized equations always have coefficients that depend on time because the fundamental nonlinear solution that determines them is not steady, but depends on time itself. Consequently, solutions do not generally have simple exponential behavior in time, as is sometimes assumed.

Interpreting the behavior of solutions to the linearized equations requires extra care because of the inherent time dependence in the coefficients. Although the method of "frozen coefficients" has often been used, apparently with some success, its reliability is truly unknown. Counterexamples in the literature demonstrate that the method can give completely wrong results for some systems of equations. One such example is quoted in Equations (6.1) and (6.2). As an example of a different nature in this chapter, Section 6.2.2, it has been shown that a strong initial boundary layer in time forms for linearized work hardening systems. The initial behavior in the boundary layer is in no way indicative of the true response of the physical system at longer times. In fact, although a perturbation in strain rate rapidly tends toward zero initially, after emerging from the boundary layer, it may grow steadily as indicated by the last of Equations (6.35) and (6.41).

An examination of a series of special cases with and without heat conduction and/or work hardening shows the generally stabilizing influence of the former and the partially stabilizing influence of the latter. Numerical simulation, discussed in Section 6.2.5, shows that the qualitative behavior of the nonlinear system of equations is preserved by the asymptotic solutions to the linearized system. Furthermore, the asymptotic solutions reveal the importance of thermal sensitivity, strain rate sensitivity, and their ratio in determining many of the qualitative features of material instability. They also reveal that only a properly scaled, composite defect has an influence on the solution outside the initial boundary layer. Furthermore, for small defects the effective defect is the sum of the scaled component defects.

Even though the perturbation in stress must vanish identically according to the linearized equations, it is still possible to make an estimate of the critical strain at which severe localization occurs and at which the actual stress collapses. This is done in Section 6.2.6 by finding the point at which the perturbation in strain rate equals the nominal applied rate. The value found in this way cannot be precisely accurate, of course, but it does suggest that there should be a simple relationship between an initial defect and the critical strain when both are properly scaled. Furthermore, the scale factors for both defect and strain are determined by the analysis. The scale factor for the critical strain may be associated with a susceptibility to adiabatic shear, but the scaled defect also has a strong role in determining the actual critical strain. Examination of the case of perfect plasticity with thermal conductivity reveals a multiplying factor, shown in Equation (6.48), which may be used quantitatively to estimate the delaying influence of heat conduction when the strain rate is less than a few thousand per second.

A relationship between scaled defect and critical strain for work hardening materials may also be found by determining the time in the linearized solution when the perturbed strain rate equals the applied strain rate. In this case the method delivers Equation (6.53), which appears to be most accurate near peak homogeneous stress. Numerical experimentation has shown that another relationship, Equation (6.55), which is similar to the one for perfectly plastic materials, gives a much better estimate at points somewhat removed from peak stress. The relationships suggested by the linearized theory and by numerical experimentation are actually tangent to each other at peak homogeneous stress.

Finally, in Section 6.3 after the determination of growth rates, it was argued that important features and patterns may be revealed such as the minimum likely spacing between nucleating bands, the applied strain rate that will minimize the critical strain in a torsional Kolsky bar experiment, or the maximum gage length that will favor formation of only one band in an experiment.

7

One-dimensional problems, part III: Nonlinear solutions

As in the previous chapter, it is easier to work with the quasi-static equations than with the full dynamic equations. These are appropriate for examining the formation of shear bands in a finite strip of material, provided the nondimensional density is not too large. In addition, suppose that the equations are to be applied to the torsional Kolsky bar experiment on a thin-walled tube where the relative wall thickness is $\ell(y)$, which has an average value of one. The modified equations become

$$
\begin{aligned}
(\ell s)_y &= 0, & &\text{momentum;} \\
\ell\theta_t &= k(\ell\theta_y)_y + \ell s\dot{\gamma}, & &\text{energy;} \\
\kappa_t &= M(\kappa, \theta)s\dot{\gamma}, & &\text{work hardening;} \\
s &= F(\kappa, \theta, \dot{\gamma}), & &\text{flow rule.}
\end{aligned}
\tag{7.1}
$$

Here $\dot{\gamma} = v_y$ is the nondimensional strain rate. Recall also that the nondimensional density is given by $\rho = \bar{\rho}(\dot{\gamma}_0 H)^2/s_0$. With $\bar{\rho} \approx 10^4\,\text{kg/m}^3$, $\dot{\gamma}_0 H \approx 3\,\text{m/s}$, and $s_0 \approx 1\,\text{GPa}$, we have $\rho \approx 10^{-4}$, so the inertia term is relatively very small and may be justifiably ignored.

7.1 Adiabatic Cases; $k = 0$

In the absence of heat conduction, solutions to the nonlinear equations (7.1) have been given by Molinari and Clifton (1987) and by Wright (1990a, 1990b).

7.1.1 Stress Boundary Condition

If stress is specified on the boundary, then $\ell s = f(t)$ is a known function. The energy and work hardening equations may be combined to give

150

$d\kappa = M(\kappa, \theta)\,d\theta$, with initial conditions $\kappa = 1 + \delta\kappa(y)$, $\theta = \delta\theta(y)$ where $\delta\kappa$ and $\delta\theta$ indicate possible variations from uniform initial conditions. The solution to the ODE may be written $\kappa = \hat{\kappa}(\theta; y)$. Inversion of the flow rule may be written $\dot{\gamma} = \Gamma(s, \kappa, \theta)$, or because ℓs and κ are known functions, we may write $\dot{\gamma} = \Gamma[f(t)/\ell(y), \hat{\kappa}(\theta; y), \theta] = \hat{\Gamma}(\theta, t; y)$. Now the energy equation takes the form of a nonautonomous ODE with y as a parameter:

$$\ell(y)\frac{d\theta}{dt} = f(t)\hat{\Gamma}(\theta, t; y). \tag{7.2}$$

The initial condition is $\theta(0, y) = \delta\theta(y)$.

If the applied load is a constant, $\ell(y)s_0(y) = 1$ after nondimensionalization, the ODE is autonomous and may be written $d\theta/\hat{\Gamma}(\theta, y) = dt/\ell(y)$. Integration essentially yields the solution given by Molinari and Clifton (1987):

$$\ell(y)\int_{\theta_0(y)}^{\theta(y,t)} \frac{d\xi}{\hat{\Gamma}(\xi; y)} = t, \tag{7.3}$$

which gives the solution for $\theta(y, t)$ in implicit form.

7.1.2 Velocity Boundary Condition; Wright (1990a, 1990b)

If velocity is specified on the boundary, and if in addition the flow law has the form of a power law $s = F(\kappa, \theta)v_y^m$, then the solution of (7.1) in implicit form may be reduced exactly to a sequence of quadratures. As before, the work hardening may be expressed in terms of a function of temperature, $\kappa = \hat{\kappa}(\theta)$, where the y dependence has been suppressed. With the definition $G(\theta) = F[\hat{\kappa}(\theta), \theta]$ the flow law now has the form

$$s = G(\theta)v_y^m. \tag{7.4}$$

During homogeneous deformation the nondimensional strain rate is given by $v_y = 1$, so the function $G(\theta)$ represents the homogeneous, adiabatic response for stress. Because of work hardening, G increases initially, but because of thermal softening, it reaches a maximum and then decreases. The energy equation may be rewritten as

$$\frac{d\theta}{dt} = s\dot{\gamma} = s(s/G)^{1/m} = s^{(1+m)/m}G^{-1/m}(\theta),$$
$$= (\ell s)^{(1+m)/m}G^{-1/m}\ell^{-(1+m)/m}. \tag{7.5}$$

Because the product ℓs depends only on time, according to the first equation of (7.1), a new quasi-time, T, may be defined by

$$\frac{dT}{dt} = (\ell s)^{(1+m)/m}, \quad T(0) = 0, \tag{7.6}$$

which reduces (7.5) to

$$G^{1/m}(\theta)\,d\theta = \ell^{-(1+m)/m}(y)\,dT.\tag{7.7}$$

Define a monotonically increasing (hence invertible) function H as

$$H(\theta) = \int_0^{\theta} G^{1/m}(\xi)\,d\xi.\tag{7.8}$$

Then after noting that the y dependence in Equation (7.7) is only parametric, one may write the solution to (7.7) as

$$H(\theta) = H[\theta_0(y)] + T\ell^{-(1+m)/m}.\tag{7.9}$$

Equation (7.9) gives the temperature in implicit form as a function of the spatial coordinate and the quasi-time, $\theta = \hat{\theta}(y, T)$. The stress may be found by integrating the strain rate over the interval, $\int_0^1 v_y\,dy = 1 = \int_0^1 (s/G)^{1/m}\,dy = (\ell s)^{1/m}\int_0^1 (\ell G)^{-1/m}\,dy$ or

$$(\ell s)^{-1/m} = \int_0^1 (\ell G)^{-1/m}\,dy.\tag{7.10}$$

Finally, the real time may be recovered by integrating (7.6):

$$t = \int_0^T (\ell s)^{-(1+m)/m}\,dT = \int_0^T \left[\int_0^1 (\ell G)^{-1/m}dy\right]^{1+m} dT.\tag{7.11}$$

Once $G(\theta)$ has been found, the sequence of quadratures given by Equations (7.9), (7.10), and (7.11) give a complete solution to the problem in implicit and parametric form.

7.1.3 Graphical Interpretation of Solutions

These equations have a simple graphical interpretation that shows how localization occurs. Figure 7.1 shows the response of a typical nonwork hardening material. The softening function $g(\theta)$, the exponentiated function $g^{1/m}(\theta)$, and the integral $H(\theta) = \int_0^{\theta} g^{1/m}(\xi)\,d\xi$ are all shown. The function g decreases slowly, but because the power $1/m$ is typically much greater than one (m is typically less than 0.03 for metals), the exponentiated function decreases rapidly toward zero. As a consequence, its integral rises rapidly from zero but approaches a plateau.

In the case of a work hardening material the adiabatic response $G(\theta)$ does not decrease initially, but rises first toward a maximum, and then decreases slowly. Now the exponentiated function $G^{1/m}(\theta)$ has a sharp maximum, as shown in Figure 7.2, and the integral $H(\theta)$ has an inflexion at the maximum of G before

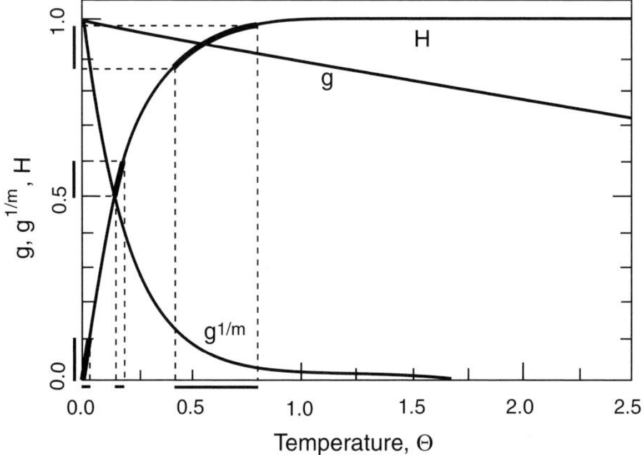

Figure 7.1. Graphical interpretation of localization for a temperature perturbation in a nonwork hardening material without thermal conduction. The figure shows the adiabatic softening curve, g, the exponentiated curve, $g^{1/m}$, and the integral of the exponentiated curve, H. With $\ell = 1$, the implicit solution shows that $H[\theta(y, t)] = H[\theta_0(y)] + T(t)$. As time increases, the distribution of responses, H, translates up the vertical axis, as shown by the dark lines. The dotted lines show the limits of the map back to temperature. Because the H response curve turns to the right, the corresponding distribution of temperature broadens substantially. (Reproduced from Wright 1990a, with permission from Elsevier Science.)

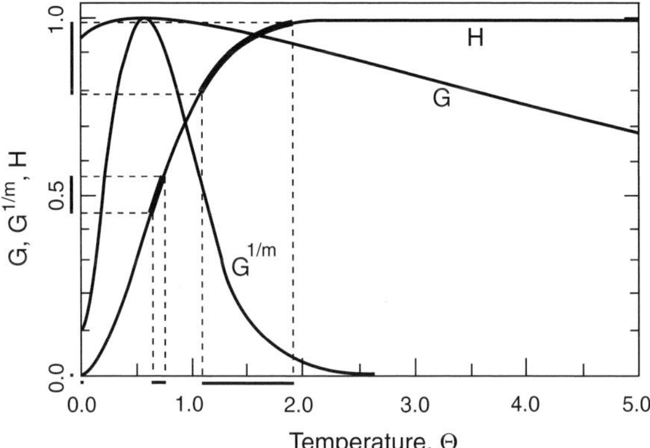

Figure 7.2. Graphical interpretation of localization for a mechanical perturbation in a work hardening material without thermal conduction. The adiabatic response curve has a maximum, and the exponentiated curve has a sharp maximum at the same temperature. The integrated response curve, H, still shows an asymptotic limit for large temperature. For a finite mechanical perturbation, $\ell(y) = 1 - \varepsilon(y)$, the implicit solution is $H[\theta(y, t)] = T(t)\ell^{-(1+m)/m}(y)$. As time increases, the distribution of H becomes wider and translates up the vertical axis, as shown by the dark lines. The dotted lines show the limits of the map back to temperature. Because the H response curve turns to the right, the corresponding distribution of temperature broadens substantially. (Reproduced from Wright 1990a, with permission from Elsevier Science.)

rising to a plateau. (All curves in the two figures have been rescaled so that each has a maximum of 1.0.)

Now suppose that the wall of the sheared specimen has the same thickness all along its length so that the relative wall thickness is exactly one, that is $\ell(y) \equiv 1$, but that the initial temperature is nonuniform, say $\theta(y, 0) = \theta_0(y)$, where θ_0 is an even function. This initial distribution maps into another distribution $H[\theta_0(y)]$. For the perfectly plastic case the range of these two initial functions is shown in Figure 7.1 by the dark lines near the origin. As time increases, the quasi-time $T(t)$ increases too because the right-hand side of (7.6) is positive. The solution from (7.9) with $\ell = 1$ is $H(\theta) = H(\theta_0) + T$ and shows that for each location y, the function H simply translates up the vertical axis by an amount T, as shown by the dark vertical lines near 0.5 and 1.0. Because the curve $H(\theta)$ turns to the right, the inverse of H maps the temperature back onto an expanding region along the horizontal axis. Finally, when the peak value of H approaches the plateau, the peak temperature becomes very large, although the minimum temperature remains moderate. This exaggeration of the upper temperatures corresponds to localization.

The graphical interpretation for a temperature perturbation in a work hardening material is similar except that the temperature distribution must first compress before expanding. This happens because the H curve first curves up and then passes through an inflexion before turning to the right and approaching a plateau. The process may easily be visualized by referring to Figure 7.2, although it is not shown, and by imagining the inverse map as the initial distribution of H slides up the vertical axis without distortion.

The graphical interpretation for a mechanical imperfection is somewhat different. Referring to Equation (7.9), with no temperature perturbation we have $H[\theta_0(y)] = H(0) = 0$, and the implicit solution is just $H(\theta) = T\ell^{-(1+m)/m}$. The exponent on ℓ has a small denominator, so even a small decrease in thickness has a major effect. For example, if $0.99 \le \ell(y) \le 1.01$, and $m = 0.02$, the exponentiated term varies over $1.67 \ge \ell^{-51} \ge 0.60$. As time increases, the right side of (7.9) increases linearly with $T(t)$, and the midpoint of the pattern moves up the vertical axis as the width of the pattern expands. This is shown in Figure 7.2. Eventually the top of the pattern, which corresponds to the thinnest section, moves onto the upper plateau, and the temperature at that same point increases rapidly, as indicated by the inverse mapping from H to θ.

Still a third variation of the graphical interpretation is necessary if there is an initial perturbation in strength, but none in temperature or thickness, so that $\kappa(y, t = 0) = \hat{\kappa}(\theta = 0; y)$. Now there is a family of H curves through the origin corresponding to each value of y, and the solution may be written $H(\theta; y) = T$. As shown in Figure 7.3, the origin corresponds to the initial moment when all

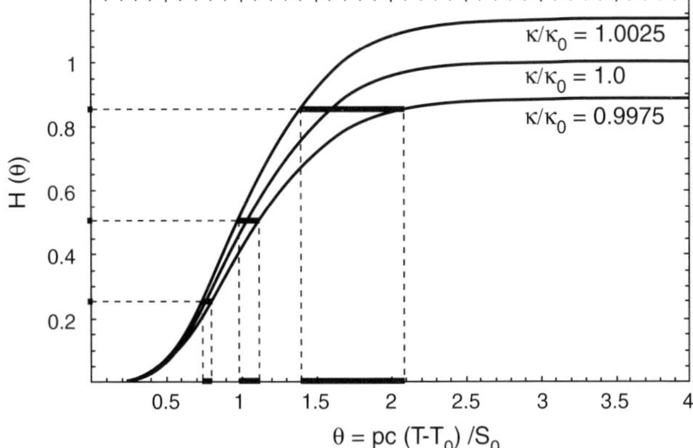

Figure 7.3. Graphical interpretation of localization for a perturbation in the material strength of a work hardening material without thermal conduction. Each point in the material has its own integrated response curve, $H(\theta; y)$. Because of the weak strain rate sensitivity, a variation in strength of only $\pm 1/4\%$ has a strong effect on the response curves. In this case the solution may be written $H[\theta(y, t); y] = T(t)$, so the distribution of H is a single point with a value that increases with time. The dotted lines show the map back to temperature through a family of H curves. Because the response curves turn to the right, the corresponding distribution of temperature broadens substantially.

points have the same temperature. As T increases, each material point has the same value of H, but the map back to temperature will begin to show a spread as the various H curves begin to diverge from one another. Eventually, as the lowest H curve turns toward the horizontal, the corresponding temperature will increase rapidly and diverge from adjacent points.

In each of the cases described above, heat conduction has been ignored. As a consequence, the solution can only be extended until extreme temperature in the center of the band and a singularity in strain rate occur. This may be seen graphically in Figures 7.1, 7.2, and 7.3. According to Equation (7.9), as the pseudo-time, T, increases monotonically with the real time, t, the distribution $H(\theta)$ translates up the vertical axis until its upper end reaches the asymptotic limit of the curve. At that point it can translate no further and the mathematical solution becomes singular. This will occur no matter what the value of the strain rate sensitivity as long as the thermal conductivity vanishes.

In spite of the three different graphical interpretations, all three types of perturbation (temperature, strength, and thickness) can be scaled in such a way that weak perturbations in one are indistinguishable from any other. This has already been suggested in Chapter 6, where it was shown that the composite defect, $\tilde{\lambda}_0(y) = \tilde{\kappa}_0(y) - M(0)\tilde{\theta}_0(y)$, determined the linearized solution (after startup) rather than the individual perturbations in strength or temperature.

In addition, fully nonlinear calculations showed that the time of severe local-ization depends only on $\tilde{\lambda}(0)$, rather than on its separate parts, at least for small perturbations of only 1–2%.

The preceding qualitative discussion shows that a flat region in the H func-tion, as defined by Equation (7.8), favors localization. For small values of m the plateau may be well developed at modest values of temperature. This cor-responds to the notion that temperatures immediately before localization are not necessarily very high, although once severe localization has developed, the central temperature may approach a substantial fraction of the melting temperature. The largest temperature change occurs during the actual, rapid transition to full localization. It may be recalled from Equation (5.2) that the scale for temperature is $s_0/\rho c$. For example, for a typical strong steel with $s_0 \approx 5$ to 8×10^8 Pa, $\rho \approx 8 \times 10^3$ kg/m^3, and $c \approx 0.5 \times 10^3$ J/kg K, the temper-ature scale is only 125–200 °C. Often localization occurs when the nondimen-sional temperature is in the range of 0.5 to 1.0, which is to say at a temperature rise of no more than one unit of the temperature scale. During localization the rise may be as much as an order of magnitude more, but even then the tem-perature may remain well below melting, although other transitions or phase changes may occur. In a typical event the full loading and unloading cycle for a particular material element may occur within a few hundred microseconds or less. At these high rates, the dynamics of a phase change should probably be included in the modeling, rather than just the transition temperature, if the extent of the change is to be well represented. Models of this complexity have not been explored.

7.2 Finite Thermal Conductivity; $k \neq 0$

Finite thermal conductivity retards the transition to localization and removes the mathematical singularity at localization. In the core of a fully developed shear band, heat conduction and heat production from plastic work are nearly balanced, as will be seen later. The addition of heat conduction is a significant complication for analysis, but some approximate results are still possible.

7.2.1 An Exact Solution

As far as the author is aware, this solution, given by Wright (1990b), is the only known exact solution to the shearing equations when finite thermal conductivity is included. The example is perhaps overly simplistic and somewhat degenerate, but it does demonstrate that strain softening is not enough by itself to cause localization.

Consider the following equations for shearing of a particular rigid-plastic material with no work hardening. The equations are in the nondimensional form for quasi-static loading:

$$
\begin{aligned}
s_y &= 0, & &\text{momentum;} \\
\theta_t &= k\theta_{yy} + s\dot{\gamma}, & &\text{energy;} \\
s &= (1 + a\theta/m)^{-m}\dot{\gamma}^m, & &\text{flow law.}
\end{aligned}
\tag{7.12}
$$

Assume that the material has insulated boundaries, $\theta_y(0, t) = \theta_y(1, t) = 0$, and that the initial condition, $\theta(y, 0) = \theta_0(y)$, is arbitrary except that the average initial temperature is zero; $\int_0^1 \theta_0(y)dy = 0$. Because of the rigid-plastic assumption, the plastic strain rate equals the velocity gradient, $\dot{\gamma} = v_y$. Because of the nondimensionalization, $v(1, t) = 1$, and the strain rate has an average value over the unit interval of one, $\int_0^1 \dot{\gamma}dy = \int_0^1 v_y dy = 1$. When the temperature is zero, the initial rate of thermal softening is $\partial s/\partial\theta = -a$, so a is the usual softening coefficient, and m is the strain rate sensitivity. Finally, because of the quasi-static assumption, the stress depends only on time, and therefore, after substitution from the third equation of (7.12), the energy equation may be rewritten as a linear PDE with a time-dependent coefficient:

$$
(1 + a\theta/m)_t = k(1 + a\theta/m)_{yy} + (a/m)s^{(1+m)/m}(1 + a\theta/m). \tag{7.13}
$$

Let us now write the dependent variable in (7.13) as the product of a function of time and a function that satisfies a simple diffusion equation; that is,

$$
\begin{aligned}
(1 + a\theta/m) &= f(t)\psi(y, t), \\
\psi_t &= k\psi_{yy},
\end{aligned}
\tag{7.14}
$$

where

$$
\psi(y, 0) = 1 + (a/m)\theta_0(y), \quad \psi_y(0, t) = \psi_y(1, t) = 0.
$$

The initial and boundary conditions on ψ ensure that the initial and boundary conditions on θ will be met, provided that the initial condition $f(0) = 1$ is satisfied. The energy equation (7.13) is now reduced to

$$
\dot{f} = (a/m)s^{(1+m)/m} f. \tag{7.15}
$$

Both f and s are only functions of time, and both are still unknown. However, because the average of the strain rate is one, the flow law yields $\int_0^1 \dot{\gamma}dy = 1 = s^{1/m}\int_0^1 (1 + a\theta/m)\,dy$. Furthermore, because ψ satisfies a diffusion equation with insulated boundaries, the average of ψ is a constant for all times. Thus, $\int_0^1 \psi\,dy = \int_0^1 \psi(y, 0)\,dy = 1$, because the average initial temperature vanishes. Finally, by averaging the first equation of (7.14), we obtain

$\int_0^1 (1 + a\theta/m)\, dy = f$. As a consequence, s and f are related by $1 = s^{1/m} f$ and (7.15) becomes

$$\dot{f} = (a/m) f^{-m}. \qquad (7.16)$$

Equation (7.16) has the solution $f = \{1 + a[(1 + m)/m]t\}^{1/(1+m)}$, and the solution to the problem is

$$s = \left(1 + \frac{1 + m}{m} at\right)^{-m/(1+m)},$$

$$\theta = \frac{m}{a}\left[\left(1 + \frac{1 + m}{m} at\right)^{1/(1+m)} \psi - 1\right], \qquad (7.17)$$

$$\dot{\gamma} = \psi = 1 + \sum_{n=1}^{\infty} \psi_n e^{-k(n\pi)^2 t} \cos n\pi y.$$

This solution also includes both the homogeneous case, because $\psi(y, t) = \psi(y, 0) = 1$ when $\theta_0 \equiv 0$, and the non-heat-conducting case where $k = 0$. The stress decays weakly in exactly the same fashion for all circumstances. There is never any localization, even in the absence of heat conduction, but in that case strain rate never decays to a uniform value either.

7.2.2 Approximate Solution With No Work Hardening

This solution was given by Wright (1990a). In this case, the nondimensional flow law is assumed to be

$$s = g(\theta)\dot{\gamma}^m. \qquad (7.18)$$

Because s depends only on time in the quasi-static case, it is possible, as was done previously, to define a new temporal variable T and a new thermal variable H by

$$\frac{dT}{dt} = s^{(1+m)/m}, \quad T(0) = 0,$$

$$H(\theta) = \int_0^\theta g^{1/m}(\xi)\, d\xi. \qquad (7.19)$$

In these variables the energy equation transforms into

$$H_T = 1 + k\frac{dt}{dT}\left(H_{yy} - \frac{H_{\theta\theta}}{H_\theta^2} H_y^2\right). \qquad (7.20)$$

If the second term in the parentheses is ignored, the equation may be rewritten as $(H - T)_t = k(H - T)_{yy}$, to which the solution is

$$H(\theta) = T + \psi(y, t),$$
$$\psi_t = k\psi_{yy}, \quad \psi(y, 0) = H[\theta_0(y)], \quad \psi_y(0, t) = \psi_y(1, t) = 0. \tag{7.21}$$

The temperature, $\theta(y, t, T)$, is given in implicit form in (7.21). As before, an expression for the stress is found by integrating the strain rate from the flow law over the unit interval

$$\int_0^1 \dot\gamma \, dy = \int_0^1 v_y \, dy = 1 = s^{1/m}(t, T) \int_0^1 g^{-1/m}[\theta(y, t, T)] \, dy. \tag{7.22}$$

Finally, the time is recovered by integrating the nonautonomous ODE:

$$\frac{dT}{dt} = s^{(1+m)/m}(t, T) = \left[\int_0^1 g^{-1/m}[\theta(y, t, T)] \, dy \right]^{-(1+m)}, \quad T(0) = 0. \tag{7.23}$$

Equations (7.21), (7.22), and (7.23) give the solution to the shear band equations in parametric and implicit form. They can be used as a basis for calculations, as in Wright (1990a), where they were shown to give accurate estimates of the critical strain at the onset of severe localization over a significant range, as shown in Figure 7.4. The figure shows that accurate estimates were obtained over two orders of magnitude of the nominal strain rate. The approximation failed when the nominal rate was too close to the stable limit and when it was

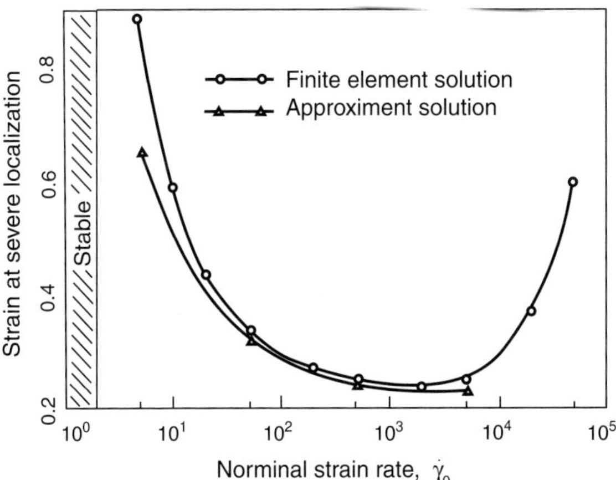

Figure 7.4. Equations (7.21), (7.22), and (7.23) lead to an effective approximation for the U curve over several decades of nominal strain rate. The approximation fails both near the lower limit where heat conduction provides absolute stability and to the right of the minimum where inertia becomes important. (Reproduced from Wright 1990a, with permission from Elsevier Science.)

significantly larger than the rate at the minimum of the U curve. The solution does not give an effective approximation for the fully formed shear band because of the terms that were ignored in (7.20).

With thermal conductivity set equal to zero, the minimum critical strain can be estimated effectively. Formula (6.67) then gives the nominal strain rate at the minimum of the U curve in Figure 7.4.

7.2.3 Further Approximations and Qualitative Interpretation

The greatest usefulness of the approximate solution, however, lies in qualitative interpretation and as a basis for further approximations, as was also discussed in Wright (1990a). Equation (7.21) is similar to (7.9) with $\ell = 1$, for example, and the graphical interpretation is similar, too. At $t = 0$ we have $\psi(y, 0) = H[\theta_0(y)]$, so the two solutions are initially identical, as they should be. As time increases, the solution $H(\theta)$ translates up the vertical axis by an amount T in both cases, but in one case the function $H[\theta_0(y)]$ remains fixed, whereas in the other it is only the initial condition for a simple diffusion problem. Diffusion has the effect of narrowing the range of ψ so that the top of the range does not run onto the plateau of the H curve as soon as in the nondiffusing case. The obvious effect is to delay the rapid separation of the highest and lowest temperatures. In other words the amplification promoted by the plateau is partially offset by the diffusion.

Further progress may be made by approximating the H function. First, define a new thermal variable $z \equiv -\ln g(\theta)$ or $g(\theta) = e^{-z}$. Then use Laplace's method (e.g., see Bender and Orszag 1978) on the definition of the H function to derive an asymptotic expression for the integral.

$$H = \int_0^\theta g^{1/m}(\xi)\,d\xi = \int_0^z e^{-z/m}\frac{d\theta}{dz'}\,dz'. \tag{7.24}$$

Assume that the derivative may be written as a power series, $d\theta/dz = a_0 + a_1 z + a_2 z^2 + \cdots$. Then the coefficients may be found by differentiating the definition for z and evaluating at $z = 0$ to get

$$a_0 = \frac{d\theta(0)}{dz} = a^{-1}, \quad a_1 = \frac{d^2\theta(0)}{dz^2} = a^{-1}\left[-1 + \frac{g''(0)}{a^2}\right], \tag{7.25}$$

and so on, where $a = -dg/d\theta$ evaluated at $\theta = 0$. Term by term evaluation of (7.24) leads to

$$H = m(a_0 + a_1 m + 2a_2 m^2 + \cdots)(1 - e^{-z/m})$$
$$-mze^{-z/m}(a_1 + 2ma_2 + \cdots) + mz^2 e^{-z/m}(a_2 + \cdots) + \mathcal{O}(z^n e^{-z/m}). \tag{7.26}$$

Because m is small, the terms in $mz^n e^{-z/m}$ will be ignored for all $n \geq 1$. Then a good approximation for H is just the simple expression

$$H = (mC/a)\left(1 - g^{1/m}\right),\tag{7.27}$$

where C is a constant chosen so that H takes on the correct limit for increasing θ. If $g = e^{-a\theta}$, the expression is exact and $C = 1$. For other choices of g the constant may be expressed as $C = 1 + \mathcal{O}(m)$.

Approximate derivatives of H follow from the definition together with (7.27):

$$H'(\theta) \equiv g^{1/m}(\theta) = 1 - \frac{H}{mC/a}, \qquad H''(\theta) = -\frac{a}{mC}\left(1 - \frac{H}{mC/a}\right).\tag{7.28}$$

A simple linear rescaling of the variables

$$\hat{H} = \frac{H}{mC/a}, \quad \hat{T} = \frac{T}{mC/a}, \quad \hat{t} = \frac{t}{mC/a}, \quad \hat{k} = \frac{mC}{a}k\tag{7.29}$$

leads to a universal approximate equation for $\hat{H}$, no matter what the softening function may be:

$$\hat{H}_{\hat{t}} = 1 + \hat{k}\frac{d\hat{t}}{d\hat{T}}\left[\hat{H}_{yy} - \frac{(\hat{H}_y)^2}{1 - \hat{H}}\right].\tag{7.30}$$

From (7.27) it is known that to a good approximation $g^{-1/m} = (1 - \hat{H})^{-1}$, and therefore from (7.23)

$$\frac{d\hat{t}}{d\hat{T}} = \left[\int_0^1 \frac{dy}{1 - \hat{H}}\right]^{1+m}, \qquad \hat{t}(0) = 0.\tag{7.31}$$

The net result of the approximation for H and the rescaling is the recognition that the left side of the U curve in Figure 7.4 (the side where heat conduction is important but inertia is not), when properly scaled, is a universal approximation for the critical strain at a fixed defect in temperature. That is to say, because $\hat{k}^{-1} = (mCk/a)^{-1} = (\kappa_0 \bar{a}/m\bar{\rho}cC)(\bar{\rho}ch^2/\bar{k})\dot{\gamma}_0 = (\kappa_0\bar{a}h^2/m\bar{k}C)\dot{\gamma}_0$, it is proportional to the nominal strain rate, and therefore, the critical strain, which equals the critical nondimensional time, may be written as

$$\gamma_{cr} = \hat{t}_{cr} = U(\hat{k}^{-1}; m),\tag{7.32}$$

where for a given defect, the function U represents the left side of the curve in Figure 7.4 with linearly rescaled axes. In other words, the figure when properly scaled will be identical for all materials, provided only that they have the same strain rate sensitivity.

The preceding analysis shows that the dominant material characteristics affecting localization are the initial slope for thermal softening, $a = -g_\theta(0)$, and the strain rate sensitivity, m. The shape of the softening curve has only a minor effect through the constant C, which is close to one for all materials. Numerical experiments by Walter (1992) confirmed the importance of the initial rate of softening and the relative unimportance of the remainder of the softening function.

The physical reason why the initial slope is important, but not the subsequent shape, is that the temperature rise up to the point that localization is just incipient is often rather modest, and therefore a linear approximation for the softening function captures most of the effect. For example, suppose that $\hat{H} = 0.99$ so that it lies just 1% short of the plateau. Then according to (7.27), $g^{1/m} = 0.01$ or $g = (0.01)^m$. Because m is small, g is actually close to one, and its argument, $a\theta$, is close to zero. With m in the range of 0.01 to 0.03, g lies in the range of 0.95 to 0.87, and if $g = e^{-a\theta}$, the product $a\theta$ lies in the range of approximately 0.051 to 0.140. A reasonable value for a is approximately 0.1 for a high-strength steel ($\bar{\rho} \approx 7800 \text{ kg/m}^3$, $c \approx 500 \text{ J/kg K}$, and $\kappa_0 \approx 0.5 \text{ GPa}$). With these values the temperature increase that corresponds to $\hat{H} = 0.99$ is in the range of $55\,^\circ$C to $165\,^\circ$C. These increases, reached just before intense localization, are still small compared with those reached in typical, fully formed shear bands (Hartley et al. 1987). Note that the higher temperatures at incipient localization correspond to larger strain rate sensitivity.

As mentioned above, the intrinsic, parametric solution can also be used as a basis for further approximations. For example, suppose that the softening function is $g = e^{-a\theta}$ with initial temperature $\theta_0 = \varepsilon \cos \pi y$ where the amplitude is small, $\varepsilon \ll m/a$ (or $\bar{\theta} \ll 25$–$40\,^\circ$C). Then, according to (7.19), $H(\theta_0) = \int_0^{\theta_0} e^{-a\theta/m} d\theta = (m/a)(1 - e^{-a\theta_0/m}) \approx \varepsilon \cos \pi y$. Equation (7.21), after rescaling, reads $\hat{H} = \hat{T} + (a\varepsilon/m)e^{-\hat{k}\pi^2\hat{t}} \cos \pi y$, and Equation (7.23), written in the form of (7.31), reads

$$s^{-1/m} = \int_0^1 \frac{dy}{1 - \hat{H}} = \int_0^1 \frac{dy}{1 - \hat{T} - \hat{\psi}}$$

$$= \frac{1}{1 - \hat{T}} \int_0^1 \frac{dy}{1 - [(a\varepsilon/m)/(1 - \hat{T})]e^{-\hat{k}\pi^2\hat{t}} \cos \pi y} \tag{7.33}$$

$$= \left[(1 - \hat{T})^2 - (a\varepsilon/m)^2 e^{-2\hat{k}\pi^2\hat{t}}\right]^{-1/2}.$$

Equation (7.23) now reads

$$\frac{d\hat{t}}{d\hat{T}} = \left[(1 - \hat{T})^2 - (a\varepsilon/m)^2 e^{-2\hat{k}\pi^2\hat{t}}\right]^{-(1+m)/2}. \tag{7.34}$$

This is a nonautonomous ODE, which must be solved numerically in general. However, in the limit when $\hat{k} \to 0$ (which means that $\dot{\gamma}_0$ equals or exceeds the minimum in the U curve in Figure 7.4), the ODE is autonomous, and if $m \ll 1$ so that it may be ignored in the exponent, then it is clear that the maximum value for $\hat{T}$ is $1 - a\varepsilon/m$ where the right-hand side becomes singular. The ODE may be solved up to that point, and the critical value for $\hat{t}$ occurs when $\hat{T}$ reaches its limit.

$$\gamma_{\mathrm{cr}} = t_{\mathrm{cr}} = \frac{m}{a} \ln \left\{ \frac{1 + \sqrt{1 - (\varepsilon a/m)^2}}{\varepsilon a/m} \right\} \approx \frac{m}{a} \ln \frac{2}{\varepsilon a/m}. \qquad (7.35)$$

Compare this result with Equation (6.46) when $k = 0$. The present result verifies that the scaling obtained previously from linear analysis is essentially correct. Only the numerical factor of 2 in the numerator of the logarithm is missing.

The preceding result has been presented in somewhat more general form (Wright 1990a), where defects in temperature, strength, and thickness of the tube wall were all considered, as well as more general softening laws. Suppose that the wall thickness, strength, and temperature have variations:

$$\ell = 1 - \delta \cos \pi y, \quad \kappa = 1 - \lambda \cos \pi y, \quad \theta_0 = \varepsilon \cos \pi y. \qquad (7.36)$$

The approximate critical strain when all three defects are present and all three are small is

$$\gamma_{\mathrm{cr}} = \frac{mC}{a} \ln \frac{2m}{a\varepsilon + (1 + m)\delta + \lambda}. \qquad (7.37)$$

The important point to notice here is that with proper scaling the three defects are equivalent to each other. That is to say, an initial defect in any one of the three (temperature or wall thickness or strength) is predicted to give exactly the same result as an equivalent defect in any other of the three.

7.3 Multiple-Length Scales: Structure of a Fully Formed Band and the Evolution of Stress in a Special Case

A recurring question in the study of adiabatic shear bands concerns the so-called halfwidth of a band. Rather than appealing to an approximate scaling argument in an attempt to answer the question of bandwidth, an asymptotic solution of the equations should reveal the full structure, at least in a special case. Then it may be possible to understand a more general case.

In numerical solutions of a particularly simple model, Wright and Walter (1987) noticed that the late stage profile of strain rate tended toward a steady state while the stress continued to decay. The material was linearly softening and perfectly plastic with power law rate hardening, as in the following nondimensional equations, which were also used by Wright and Ockendon (1992) to examine the fully formed band.

$$s_y = 0,$$

$$\theta_t = k\theta_{yy} + sv_y, \quad \theta_y(0, t) = \theta_y(1, t) = 0, \tag{7.38}$$

$$s = (1 - a\theta)v_y^m.$$

If the strain rate is to be independent of time, a more convenient variable is $u(y)$, which is related to the strain rate by $v_y = Au^{-1/m}$. The constant A represents the strain rate in the middle of the band where $u(0) = 1$. From the third equation of (7.38), the new variable is related to temperature and stress by $u = A^m(1 - a\theta)/s$. Because u is independent of time and stress depends only on time, the quantity $1 - a\theta$ is composed of a product of functions, one of which is only time dependent and the other of which is only space dependent. From the first and second equations of (7.38),

$$k\frac{u_{yy}}{u} - aA^{1+m}u^{-(1+m)/m} = \frac{\dot{s}}{s} = -\alpha. \tag{7.39}$$

Because u depends only on y, and s depends only on t, it follows from the usual argument for separation of variables that α is a constant. Thus, an observation about a numerical solution has led to a reduction of the problem to a pair of ordinary differential equations:

$$ku_{yy} + \alpha u - aA^{1+m}u^{-1/m} = 0, \quad u_y(0) = u_y(1) = 0, \quad \dot{s} + \alpha s = 0. \tag{7.40}$$

The solution for stress is $s = s_0 e^{-\alpha t}$, but it is important to note that this also gives the temporal variation of $1 - a\theta$. The amplitude A and the eigenvalue α must be determined as part of the solution.

Because the nondimensional velocity varies between zero and one, there is a condition on u, namely

$$\int_0^1 v_y \, dy = 1 = A \int_0^1 u^{-1/m} \, dy. \tag{7.41}$$

Integration of (7.40), together with (7.41) and the boundary conditions on u, leads to another condition.

$$\int_0^1 u \, dy = \frac{aA^m}{\alpha}. \tag{7.42}$$

Equation (7.42) will be used later to evaluate A.

A first integral of (7.40) that satisfies $u(0) = 1$, $u_y(0) = 0$ is

$$u_y^2 = \lambda(1 - u^{-p}) - v(u^2 - 1), \tag{7.43}$$

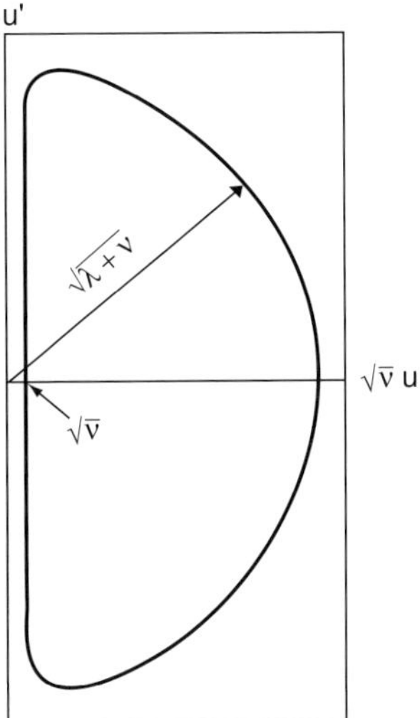

Figure 7.5. Phase portrait of Equation (7.43). Most of the curve lies on the arc of a circle of radius $\sqrt{\lambda + v}$, but as $u \to 1$, the curve plunges steeply toward zero. (Reproduced from Wright and Ockendon 1992, with permission from Elsevier Science.)

where the constants λ, p, and v are given by

$$\lambda = \frac{2am}{k(1-m)} A^{1+m}, \quad p = \frac{1-m}{m}, \quad v = \frac{\alpha}{k}. \tag{7.44}$$

Because m is small and A, the amplification of strain rate in the center of the band, is large, it can be estimated that $\lambda \gg 1$ and $p \gg 1$. Subject to verification later, it can be assumed that $v = \mathcal{O}(1)$. The phase portrait corresponding to (7.43) is shown in Figure 7.5. The right side of the figure is nearly circular with radius $\sqrt{\lambda + v}$ because $u^{-p} \ll 1$ when $u > 1$. But as $u \to 1$, the term $\lambda(1 - u^{-p})$ dominates over $v(u^2 - 1)$ as the phase curve plunges to zero at $u = 1$. Setting $u^{-p} \approx 0$ leads to an outer solution satisfying $u_y(1, t) = 0$, and setting $u^2 - 1 \approx 0$ leads to an approximate inner solution satisfying $u(0) = 1$ and $u_y(0, t) = 0$, as follows.

$$u_o = \sqrt{\frac{\lambda + v}{v}} \cos \sqrt{v}(1 - y)$$

$$u_i = 1 + \frac{2}{p} \ln \left(\cosh \frac{1}{2} p \sqrt{\lambda} y \right) \tag{7.45}$$

A matching argument leads to the conclusion that $\sqrt{v} \approx \pi/2$, so the rate of stress decay is

$$\alpha = \frac{1}{4}\pi^2 k = \frac{\pi^2 \bar{k}}{4\bar{\rho}\,c\dot{\gamma}_0 h^2} \tag{7.46}$$

The reader may refer to the original paper for further details.

In (7.45) the length scale in the argument of the inner solution is $\delta = 2/p\sqrt{\lambda}$, which gives

$$\delta = \left[\frac{2mk}{(1-m)aA^{1+m}}\right]^{1/2}. \tag{7.47}$$

This is the distance for which the argument of cosh in (7.45) is one, and consequently, from the definition of u that follows (7.38), it is the location at which the strain rate is less than the maximum rate by a factor of

$$A/v_y = u_i^{1/m}(\delta) = \{1 + [2m/(1-m)]\ln(\cosh 1)\}^{1/m}$$
$$\approx e^{[2/(1-m)]\ln(\cosh 1)} = (\cosh 1)^{2/(1-m)}. \tag{7.48}$$

For example, for $m = 0.02$, the strain rate is down by a factor of 2.4 to approximately 42% of the peak value. As can be seen in Figure 7.6, which compares the asymptotic solution with a finite-element solution, at this level the strain

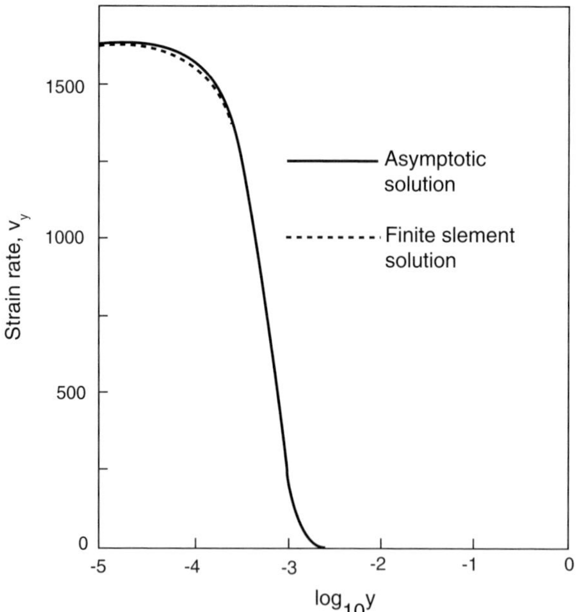

Figure 7.6. Comparison of the asymptotic and finite-element solutions. (Reproduced from Wright and Ockendon 1992, with permission from Elsevier Science.)

rate is decreasing so rapidly that this is as good a measure of the mechanical width as any.

To complete the problem, apply the normalizing condition (7.42).

$$\int_0^1 u\,dy \approx \int_0^\delta u_i\,dy + \int_\delta^1 u_o\,dy$$

$$\approx \delta + \nu^{-1/2}\sqrt{1 + (\lambda/\nu)}\,[\sin\sqrt{\nu}(1 - \delta)] \approx \nu^{-1}\sqrt{\lambda}. \tag{7.49}$$

Because the left side of (7.49) equals $a A^m/\alpha$ according to Equation (7.42), and the right side equals $k\sqrt{\lambda}/\alpha$ after using the third equation of (7.44), we have $A^m = k\sqrt{\lambda}/a$. Finally, after using the first equation of (7.44), we find that the magnification factor for the amplitude of strain rate comes out to be

$$A \approx \left(\frac{1 - m}{2}\frac{a}{mk}\right)^{1/(1-m)}. \tag{7.50}$$

A comparison of (7.50) with (7.47) shows that $A\delta = 1$, which is simple and straightforward, but its full meaning does not become clear until put back into dimensional terms by using $\dot{\gamma}_{\max}/\dot{\gamma}_0 = A$ and $\bar{\delta}/h = A^{-1}$:

$$\frac{\dot{\gamma}_{\max}}{\dot{\gamma}_0}\frac{\bar{\delta}}{h} = 1 \quad \text{or } \bar{\delta}\dot{\gamma}_{\max} = v_0, \tag{7.51}$$

where $\pm v_0$ is the velocity applied to the boundary on either side of the shear band. Equation (7.51) makes the remarkable statement that *the product of the characteristic length scale in the most rapidly varying part of the band times the maximum strain rate in the center of the band is equal to half the velocity difference across the band.*

Furthermore, when the ratio of nondimensional numbers a/k is expressed in terms of dimensional quantities,

$$\frac{a}{k} = \frac{\bar{a}\kappa_0(b\dot{\gamma}_0)^m}{\bar{\rho}c}\frac{\bar{\rho}c\dot{\gamma}_0 h^2}{\bar{k}} = \frac{\bar{a}\kappa_0 b^m \dot{\gamma}_0^{1+m} h^2}{\bar{k}}, \tag{7.52}$$

and the relation $v_0 = h\dot{\gamma}_0$ is again applied, the maximum strain rate and the characteristic length scale from (7.50) and (7.47) may be expressed in dimensional terms as

$$\dot{\gamma}_{\max} = \left(\frac{1 - m}{m}\frac{\bar{a}\kappa_0 b^m}{\bar{k}}\right)^{1/(1-m)} v_0^{2/1-m},$$

$$\bar{\delta} = \left(\frac{1 - m}{m}\frac{\bar{a}\kappa_0 b^m}{\bar{k}}\right)^{-1/(1-m)} v_0^{-(1+m)/(1-m)}. \tag{7.53}$$

It must be emphasized that the length scale just given refers to the rapid variation of velocity and strain rate across the shear band, as is obvious from the derivation. Therefore, it is a mechanical length scale and refers to the region of plastic working that is the source of heating for the adjacent material, but it is also the scale that is visually apparent in a shear band when the flow lines are clearly visible. This point will be returned to later.

Equation (7.53) indicates that, for this simple material model, the maximum strain rate in the band and the characteristic length scale in the band depend only on the imposed velocity at the edge of the band, not on the nominal strain rate or the length scale h individually. For a given material the maximum rate varies roughly as the square of the driving velocity, whereas the width decreases like the reciprocal of the velocity. The maximum rate increases nearly directly with the strength and rate of thermal softening and nearly inversely with the strain rate sensitivity and the thermal conductivity. The effect of these material properties on the width is exactly opposite to their effect on the maximum rate. Compare this behavior with that indicated in (7.75) where the scaling superficially appears to be quite different.

As an example, suppose that a steel has a strength of 0.5 Gpa, a softening rate of $1/1200$ K^{-1}, a strain rate sensitivity of 0.02, a thermal conductivity of 50 W/mK, and a driving velocity of ± 1 m/s. Then, according to (7.53), the predicted maximum strain rate is 6.4×10^5 s^{-1}. (The normalizing time b was chosen to be 10^4 s, but its exact value is of little importance because the exponent m is so small.) With the same numbers the characteristic mechanical length comes out to be 1.6 μm. By comparison, for a temperature rise of 500 K the halfwidth predicted by Bai and Dodd (1992), as discussed in Section 7.5 and given by equation (7.75), is 8.8 μm, or five and a half times as large. Although 500 K is a typical temperature rise in a shear band, the actual value varies with time. (In the linearly softening model the fully developed strain rate remains constant, but the temperature continues to rise like a slow exponential.) Although the material model used here is overly simple, it illustrates the relationship between a larger thermal length scale and the smaller mechanical length scale.

7.4 Canonical Structure of a Fully Formed Band in the General Case

In the last section the structure of a fully formed band was developed for the special case in which the material softens linearly with temperature and hardens with strain rate according to a power law. Although temperature and stress continued to evolve, the mechanical structure of the band was constant, and the innermost structure was essentially steady. Glimm, Plohr, and Sharp (1993, 1995) showed that the state of shearing is essentially steady within the core of any fully developed shear band, and for a flow law in the dimensional form

$\bar{s} = \bar{\kappa} g(\bar{\theta})(b\bar{v}_{\bar{y}})^{m}$, where $\bar{\kappa}$ is a constant, the structure of the shear band takes on a simple canonical structure that generalizes the results of the last section. Wright and Ravichandran (1997) showed that the same canonical structure actually holds for a shear band in an arbitrary material, assuming only that work hardening is essentially saturated in the center of the band. Their approach is followed in the remainder of this section.

A shear band is a one-dimensional structure in that changes perpendicular to the band are far more rapid than changes along the band. Furthermore, if the driving conditions remain fairly constant, then changes with time will also be small relative to the gradients in the perpendicular direction. Under these conditions it is reasonable to use the steady equations

$$s_y = 0, \quad (k\theta_y)_y + sv_y = 0, \tag{7.54}$$

where y is in the direction perpendicular to the band and s is the shearing traction in the direction of the particle velocity, v, along the band. In (7.54) and in the remainder of this section the equations will be in dimensional form, but for simplicity of notation, the overbars will be dropped.

The balance laws must be supplemented by constitutive equations that define a relationship among stress, strain rate, and temperature. This may be written in two forms,

$$v_y \equiv r = R(s, \theta), \quad s = S(r, \theta), \tag{7.55}$$

provided that R and S are invertible in the first variable. Because work hardening has been assumed to be fully saturated, the work hardening parameter does not appear explicitly in either form of the constitutive law. In addition, the constitutive law is used to define two auxiliary functions that are characteristic of the material, namely, the strain rate sensitivity, m, and the relative softening, $\hat{a}$.

$$m \equiv \frac{r}{S}\frac{\partial S}{\partial r} = \frac{R}{s}\frac{1}{\partial R/\partial s}, \quad \hat{a} \equiv -\frac{1}{S}\frac{\partial S}{\partial \theta} = -\frac{1}{s}\frac{\partial R/\partial \theta}{\partial R/\partial s}. \tag{7.56}$$

In (7.56) the first form for either m or $\hat{a}$ is to be regarded as a function of (r, θ), and the second as a function of (s, θ).

Equations (7.54) may be integrated in two different ways. In a straightforward manner

$$s = \text{const}, \quad k\theta_y + sv = \text{const}, \tag{7.57}$$

or after taking note that the stress is constant, use a previous result for steady shearing:

$$\frac{1}{2}(k\theta_y)^2 + s \int_{\theta_c}^{\theta} k(\theta)R(s, \theta)\,d\theta = 0. \tag{7.58}$$

In (7.58) the subscript c denotes the center of the band where the temperature gradient and velocity vanish because of symmetry. Both forms will prove useful.

It is generally true for metals that the strain rate sensitivity, which is a nondimensional number, is small and much less than one. This fact will be used to cast the integral in (7.58) into a standard form for asymptotic evaluation. Let $\lambda(s, \theta_c)$ be the strain rate in the center of the band. Next replace the function $R(s, \theta)$ by

$$R(s, \theta) = \lambda(s, \theta_c)e^{-z/m_c}, \tag{7.59}$$

where λ is the strain rate in the center of the band and m_c is a small parameter whose value will be chosen for convenience later.

Because s and θ_c are constants for any particular shear band, Equation (7.59) represents only a change of variable from θ to z. At $\theta = \theta_c$, we have $z = 0$ and $\theta < \theta_c$ corresponds to $z > 0$. The integral in (7.58) may now be written as

$$I \equiv \int_{\theta_c}^{\theta} k(\theta)R(s, \theta)\,d\theta = \lambda \int_0^z k(\theta)e^{-z/m_c}\theta_z\,dz. \tag{7.60}$$

If k is a slowly varying function of θ, and if θ_z is a slowly varying function of z, then the integral may be evaluated asymptotically by a standard method for definite integrals with a large parameter in the exponent (e.g., see Jeffreys 1962). For the moment suppose that both functions are "slowly varying" in some appropriate sense. Then expansion of both in Taylor series and term by term integration leads to a series that is asymptotic in the small parameter m_c.

$$I \sim m_c(\lambda k\theta_z)_c$$
$$\times \left\{\left(1 - e^{-z/m_c}\right) + \left[\left(\frac{k_\theta\theta_z}{k}\right)_c + \left(\frac{\theta_{zz}}{\theta_z}\right)_c\right]\right. \tag{7.61}$$
$$\left.\times \left[-ze^{-z/m_c} + m_c\left(1 - e^{-z/m_c}\right)\right] + \mathcal{O}(m_c^2)\right\}.$$

Note that because ze^{-z/m_c} has a maximum of $m_c e^{-1}$, the second set of terms in square brackets is only $\mathcal{O}(m_c)$.

Provided that the error terms in (7.61) are small enough, to leading order the integral in (7.60) may be approximated by $I = m_c(\lambda k\theta_z)_c$ $(1 - e^{-z/m_c})[1 + \mathcal{O}(m_c)]$. To interpret terms in this expression, first differentiate (7.59) to obtain $R_\theta\theta_z = -R/m_c$. Then from the definitions of the functions $\hat{a}$ and m, note that $R_\theta/R = \hat{a}/m$ and as a consequence

$$\theta_z = -\frac{m}{m_c}\frac{1}{\hat{a}}. \tag{7.62}$$

The parameter m_c, which has not yet been defined, may be identified with the strain rate sensitivity in the center of the band, as anticipated by the notation.

Therefore, we have $(\theta_z)_c = -\hat{a}_c^{-1}$. After using this result in the lowest-order expression for I, substituting into (7.58), and taking the square root, one obtains a suitable asymptotic expression for the temperature gradient as

$$k(\theta)\theta_y = \pm\sqrt{2s\left(\frac{\lambda km}{\hat{a}}\right)_c}\,(1 - e^{-z/m_c})^{1/2}[1 + \mathcal{O}(m_c)], \qquad (7.63)$$

where the exponential represents the ratio of the local strain rate to the maximum strain rate according to (7.59), and the negative sign is to be associated with positive values of y.

Equation (7.63) can be integrated with respect to y in a similar manner. First rewrite as

$$\int_0^z \frac{k(\theta)\theta_z\, dz}{(1 - e^{-z/m_c})^{1/2}} = -\left(\frac{2sk\lambda m}{\hat{a}}\right)_c^{1/2} y[1 + \mathcal{O}(m_c)]. \qquad (7.64)$$

The denominator on the left side is singular, but integrable, at the lower limit. Again it may be argued that k and θ_z are slowly varying compared with the singular term near $z = 0$, so that for lowest-order asymptotic representation they may be evaluated at the lower limit and taken outside the integral. Explicit evaluation of the integral to lowest order then gives

$$\int_0^z \frac{k(\theta)\theta_z dz}{(1 - e^{-z/m_c})^{1/2}} \sim -2\left(\frac{km}{\hat{a}}\right)_c [\cosh^{-1} e^{z/2m_c}][1 + \mathcal{O}(m_c)] \qquad (7.65)$$

The last two equations combine to give

$$\cosh^{-1} e^{z/2m_c} = y\Big/\left(\frac{2mk}{s\lambda\hat{a}}\right)_c^{1/2}[1 + \mathcal{O}(m_c)]. \qquad (7.66)$$

The denominator on the right-hand side defines the characteristic length scale, δ, for the y coordinate, namely

$$\delta = \left(\frac{2mk}{s\lambda\hat{a}}\right)_c^{1/2}[1 + \mathcal{O}(m_c)]. \qquad (7.67)$$

The distribution of strain rate, velocity, and temperature now follow in short order from inversion of (7.66), integration of the strain rate, and integration of (7.57), respectively.

$$v_y = \lambda \operatorname{sech}^2(y/\delta), \quad v = \lambda\delta\tanh(y/\delta),$$
$$\theta = \theta_c - 2(m/\hat{a})_c \ln\cosh(y/\delta)[1 + \mathcal{O}(m_c)]. \qquad (7.68)$$

This is the same result obtained by Glimm et al. (1993) for a multiplicative flow law, but here it has been obtained for a general, but fully saturated, flow law, provided that the error terms in (7.61) are small. That is to say, $m_c[(k_\theta\theta_z/k)_c + (\theta_{zz}/\theta_z)_c] = o(1)$. These will be written as two conditions:

$$m_c(k_\theta\theta_z/k)_c = o(1) \Rightarrow (k_\theta/k)_c \ll (\hat{a}/m)_c = (R_\theta/R)_c,$$

$$(m\theta_{zz}/\theta_z)_c = o(1) \Rightarrow -\left[\frac{d}{d\theta}\left(\frac{m}{\hat{a}}\right)\right]_c = o(1). \tag{7.69}$$

It will be assumed that the first equation of (7.69) holds. It is a relatively mild condition because with typical values of $\hat{a}$ at $\mathcal{O}(10^{-3})\,°\text{C}^{-1}$ and typical values of m at $\mathcal{O}(10^{-2})$, dimensional values of $\hat{a}/m$ may be expected to be in the neighborhood of $\mathcal{O}(10^{-1})\,°\text{C}^{-1}$. Thus, if the change of thermal conductivity per degree Celsius is much less than 10% of its present value, then the first equation of (7.69) will be satisfied.

The second equation of (7.69) depends on the flow law and must be verified in individual cases. However, once it has been verified, then the results in (7.67) and (7.68) are canonical no matter what the detailed form of the flow law in all other respects. Figures 7.7, 7.8, and 7.9 show the canonical forms of the strain rate, velocity, and temperature. Note that the distribution of temperature is much wider than the distribution of strain rate, as expected.

The original paper by Wright and Ravichandran (1997) gave several examples, summarized in Table 7.1, in which conditions were found that ensure the canonical structure. In every case, the strain rate sensitivity in the center of the band must be small, that is $[(r/S)(\partial S/\partial r)]_c = m_c \ll 1$. The first three cases

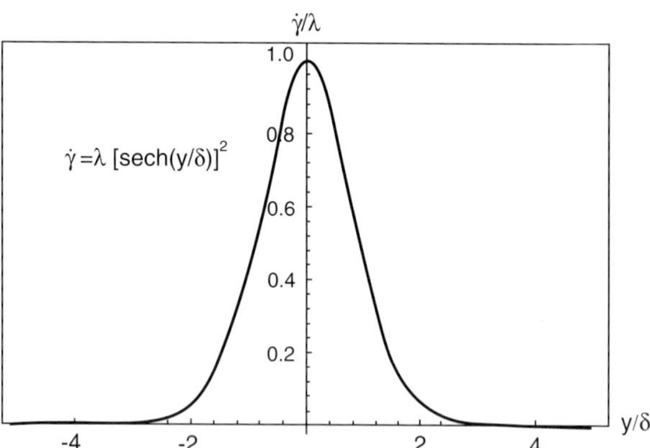

Figure 7.7. Canonical distribution of strain rate in an adiabatic shear band, as scaled by the strain rate in the center of the band, λ, and a mechanical length, δ, that is characteristic of the material in its current state. (Reproduced from Wright and Ravichandran 1992, with permission from Elsevier Science.)

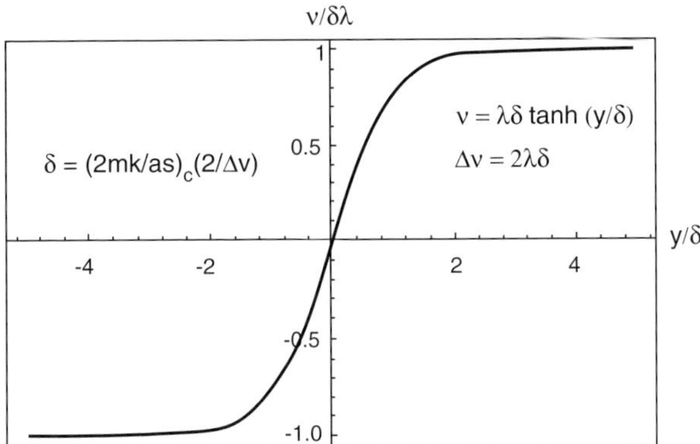

Figure 7.8. Canonical distribution of particle velocity in an adiabatic shear band relative to the center of the band, as scaled by half the jump in velocity across the band, $\Delta v/2 = \lambda\delta$, and a mechanical length, δ, that is characteristic of the material in its current state. (Reproduced from Wright and Ravichandran 1992, with permission from Elsevier Science.)

have been written so that the flow law is calibrated to the center of the band, but the last two have not.

Other relations may be found that connect the kinematic and material variables. Suppose that the stress through the band and the driving velocity exterior to the band are known, and let Δv be the jump in particle velocity across the band. Then from the three relations

$$\frac{1}{2}\Delta v = \delta\lambda, \quad \delta = \left(\frac{2mk}{s\lambda\hat{a}}\right)^{1/2}_c, \quad s = S(\lambda, \theta_c) \tag{7.70}$$

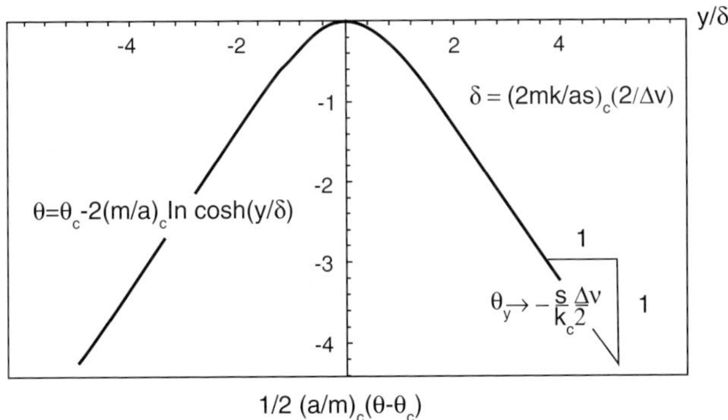

Figure 7.9. Canonical distribution of temperature in an adiabatic shear band relative to the central temperature, as scaled by the physical quantity $\rho cm/s_\theta$, and a mechanical length, δ, that is characteristic of the material in its current state. (Reproduced from Wright and Ravichandran 1992, with permission from Elsevier Science.)

Table 7.1. *Summary of conditions for asymptotic validity*

Law	Flow Law	Condition $(m\theta_{zz}/\theta_z)_c = o(1)$
1. Power Law	$s = \kappa(T/T_c)^{-v}(r/\lambda)^{\bar{m}}$	$\bar{m} \ll v$
2. Johnson–Cook	$s = s_0 + \kappa[1 + \bar{m}\,\ln(r/\lambda)]g(T - T_c)$	$\bar{m} \ll 1$
3. Zerilli–Armstrong	$s = s_0 + \kappa e^{-\beta_0(T-T_c)}(r/\lambda)^{\bar{m}+\beta_1(T-T_c)}$	$\bar{m} \ll 1$ $\beta_1 \ll \beta_0$
4. MTS model	$s = s_0 + \kappa\left[1 - (AT\ln\dot{\gamma}_0/r)^{2/3}\right]^2$	$(2m/\hat{a}T)_c \ll 1$
5. Bodner–Partom	$r = \dot{\gamma}_0 \exp\left\{-2^{-1}(\kappa^2/s^2)^{\bar{a}/T+\bar{b}}\right\}$	$(2m/\hat{a}T)_c \ll 1$

it is possible in principal to solve for the remaining variables λ, θ_c, and δ as functions of s and Δv. These three relationships come respectively from the limiting values for velocity in (7.68), Equation (7.67), and the constitutive relation (7.55) evaluated in the center of the band. The inversion may be expressed symbolically as

$$\lambda = \hat{\lambda}(s, \Delta v), \quad \theta_c = \hat{\theta}(s, \Delta v), \quad \delta = \hat{\delta}(s, \Delta v). \tag{7.71}$$

In effect, the stress, which is essentially constant through the band, and the jump in velocity across the band are regarded as the external conditions that power the band and determine the complete canonical structure. Equation (7.53) is an example of such an inversion.

The equations derived to show the canonical structure of a shear band are dynamic in nature, and they show a snapshot of the band during deformation. Although the evolving structure can be observed dynamically in a torsional Kolsky bar test, often it is only possible to observe the band in a postmortem fashion through metallurgical sections after the test is completed. For example, recall that Figure 1.3 showed a section of a plate through which a punch had been explosively driven in the vertical direction. The curved white lines that bend down toward the sheared surface are chemical inhomogeneities that were originally horizontal and parallel to the rolling plane of the plate. The approximate equation of such a line may be found by integrating the canonical velocity in (7.68) with λ and δ held constant.

$$w = (\lambda t)\delta \tanh(y/\delta) + w_0(y). \tag{7.72}$$

The product λt may be interpreted as the excess strain accumulated in the center of the band after localization occurs and the function $w_0(y)$ is the displacement field external to the band that is frozen in at localization. Because the figure indicates that w_0 should be approximately linear, Equation (7.72) may be written as

$$w = (\gamma_{\max} - \gamma_{\mathrm{crit}})\delta \tanh(y/\delta) + \gamma_{\mathrm{crit}} y \tag{7.73}$$

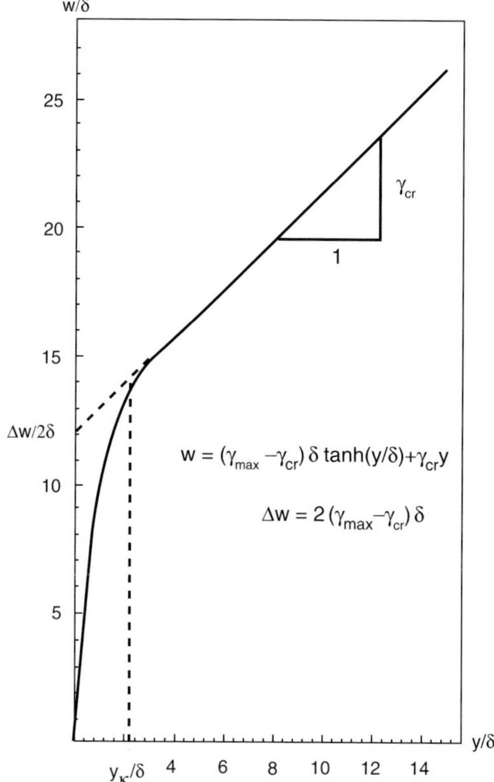

Figure 7.10. Canonical distribution of displacement across an adiabatic shear band, as scaled by a mechanical length, δ, that is characteristic of the material in its current state. Compare the theoretical distribution with the observed distribution in Figure 1.3. (Reproduced from Wright and Ravichandran 1992, with permission from Elsevier Science.)

where γ_{crit} is the critical exterior strain that is frozen into the material at the moment of localization. Figure 7.10 shows a typical case with parameters chosen to match Figure 1.3, and it shows an excellent fit to the photomicrograph when the theoretical curve is superimposed.

Strains corresponding to (7.73), which may be found by differentiation, change smoothly from γ_{max} in the center of the band to γ_{crit} exterior to the band:

$$\gamma = w_y = (\gamma_{\mathrm{max}} - \gamma_{\mathrm{crit}})\,\mathrm{sech}^2(y/\delta) + \gamma_{\mathrm{crit}}. \qquad (7.74)$$

The point of maximum curvature in Figure 7.10, denoted y_κ, may be used to estimate the value of the characteristic length, δ. Calculations show that δ/y_κ usually will lie in the range from 0.4 to 0.6. When applied to the case shown in Figure 1.3, the characteristic mechanical length is a bit shorter than

10 μm. The reader should refer to Wright and Ravichandran (1997) for further details.

7.5 Thermal and Mechanical Length Scales

A first attempt to estimate the width of a shear band was made by Dodd and Bai (1985) and reported by Bai and Dodd (1992). From the steady solution (see Chapter 5, Section 5.2) they estimated that the halfwidth in dimensional terms could be given by

$$\delta_{\mathrm{DB}} = \sqrt{\frac{\bar{k}\theta_c}{\beta\bar{s}\dot{\bar{\gamma}}_c}} = \sqrt{\frac{\bar{k}(\bar{T}_c - \bar{T}_0)}{\beta\bar{s}\dot{\bar{\gamma}}_c}} \approx \sqrt{\frac{\bar{k}\Delta\bar{\theta}}{\bar{s}\lambda}} \tag{7.75}$$

where the overbar indicates a dimensional quantity, the subscript c indicates the center of the band, and the subscript 0 indicates the ambient condition exterior to the band. Several investigators have used this estimate to compare with experimental observations, including Dodd and Bai (1985), Hartley et al. (1987), and Marchand and Duffy (1988). The theoretical and experimental values reported agree within a factor of two, and usually much less than that.

However, the decision concerning the observed feature to be measured and called the "width" must be regarded as imprecise. In some materials after etching, a shear band appears white and it is tempting to associate the width with the whitened zone. In other materials, only an intensely deformed zone can be seen, and in still others the center of the band appears to be a zone of very small, heavily refined, or possibly recrystallized grains. In particularly favorable cases it is possible to see flow lines that bend down into the band and then curve away on the other side to make an antisymmetric pattern. In all of these cases it is difficult a priori to select one width that fully characterizes the shear band. It is also difficult to estimate the conditions in the center of the band.

Formula (7.75) appears to have more to do with the thermal thickness of the band than with any mechanical length scale. The denominator, $\beta\bar{s}\dot{\bar{\gamma}}$, is the plastic power per unit volume converted to heat in a thin layer, and $\bar{k}(\bar{T}_c - \bar{T}_0)$ is a rough estimate of the thermal power conducted away per unit thickness of thermal layer.

Another length scale, δ_{mech}, which refers to the mechanical rather than the thermal structure, was found in Sections 7.3 and 7.4 by first examining a special case and then developing the general case. The mechanical length, given in Equation (7.67) for the general case, refers to the rapid change in velocity across a shear band.

A thermal length scale, δ_{th}, may also be estimated from the general asymptotic structure as follows. Figure 7.9 shows that the temperature gradient away from the core of the band is essentially constant, so a thermal width can be estimated to be $\delta_{th} = |\Delta\theta/\theta_y|$, where $\Delta\theta$ is the temperature difference between the center of the band and the adjacent ambient region, and θ_y is the asymptotic slope. From the asymptotic solution in (7.68), the absolute value of the slope is $|\theta_y| = 2m/\hat{a}_c\delta_{mech}$, where δ_{mech} is the mechanical length scale given by Equation (7.67). Consequently the estimate for δ_{th} is

$$\delta_{th} = (\hat{a}_c\Delta\theta/2m)\delta_{mech}. \tag{7.76}$$

Now by comparing Equations (7.75) and (7.67) one can see that the ratio of Dodd and Bai's thermal length to the mechanical length is just $\delta_{DB}/\delta_{mech} = \sqrt{\hat{a}_c\Delta\theta/2m} = \sqrt{\delta_{th}/\delta_{mech}}$. Rearrangement now shows that Dodd and Bai's length scale is the geometric mean of the other two,

$$\delta_{DB} = \sqrt{\delta_{th}\delta_{mech}}, \tag{7.77}$$

and therefore it lies between them, $\delta_{mech} \leq \delta_{DB} \leq \delta_{th}$.

7.6 DiLellio and Olmstead's Theory of Shear Band Evolution

In a series of papers, DiLellio and Olmstead (1997a, 1997b, 1998) developed a one-dimensional theory of shear band evolution for steady shearing of a fixed slab of material. The material model includes elasticity, exponential thermal softening (in the first and third papers), and power law rate hardening, but not work hardening, and the acceleration term is retained on the right-hand side of the momentum equation. The main feature of these analyses is the use of boundary layer theory to treat the shear band as a line discontinuity that carries the local, boundary layer structure, which in turn connects to a far field and the external boundary conditions. The strain rate sensitivity is used as the small parameter to generate a singular perturbation analysis. When the nondimensionalizing length scale is very small compared with the physical dimension of the slab, then the final result is an integral equation, coupled to an algebraic equation, with the stress and temperature in the center of the band as the two unknown functions of time. Numerical procedures are required to solve these coupled equations, but when that is done, the whole history becomes known from slow early growth, through the rapid transition to full localization, and on into a postlocalization regime.

As a byproduct, a time-dependent length scale is found, which is proportional to the mechanical length scale that is associated with the transition across the boundary layer, as expressed in Equation (7.70). Specifically, $\delta_{mech} = \ell\Delta(t)$, where ℓ is a characteristic length scale, and $\Delta(t)$ is a nondimensional measure of the halfwidth of the mechanical transition [as given by Equation (53) in their paper]. Before full localization the length scale is relatively large, but during full localization it decreases rapidly by approximately two orders of magnitude, then recovers by approximately one order of magnitude, and finally grows at a moderate rate to the end of the calculation.

There are two key features in their analysis. First, they use a nondimensionalization that differs significantly from the scheme used both in Chapter 6 and in the previous parts of this chapter. By letting the basic length scale be determined by diffusion, rather than by the dimensions of the sample, their scheme allows them to retain both elasticity and heat conduction. The boundary is then commonly at an effectively infinite distance from the shear band so that the exact thermal boundary condition becomes of little consequence. Second, they develop a perturbation scheme that assumes that the leading scaling for strain rate is order $\mathcal{O}(m^{-1})$. Then with the usual stretching transformation $\xi = y/m$, they find that the inner solution has essentially the canonical structure outlined in Section 7.4, but with the stress, temperature, and strain rate in the center of the band being functions of time (as yet unknown) rather than steady.

In the outer region the stress and temperature satisfy the simple momentum and diffusion equations in the nondimensional form:

$$s_t(y, t) = \frac{1}{\rho} \int_0^t s_{yy}(y, t')\, dt' + \frac{d}{dy} \bar{v}_0(y),$$

$$\theta_t(y, t) = \theta_{yy}(y, t), \quad 0 < y < L/\ell, \ t > 0. \tag{7.78}$$

Finally, the outer limit of the inner solution generates the boundary conditions for the outer solution at the left end of the interval while the original boundary conditions still apply to the other end of the interval. After making use of a Green's function, they arrive at two coupled integral equations for stress and temperature in the center of the developing shear band. In the limit that $L/\ell \to \infty$, these reduce to

$$s = -\left(\frac{2m\rho}{\lambda a}\right)^{1/2} \left\{ \left[\frac{\dot{\gamma}(s, \theta)}{s}\right]^{1/2} - \left[\frac{\dot{\gamma}(\bar{s}_0, \bar{\theta}_0)}{\bar{s}_0}\right]^{1/2} \right\} + \bar{s}_0 + \sqrt{\rho}\, \bar{v}_0 \left(\frac{t}{\sqrt{\rho}}\right),$$

$$\theta = 2\left(\frac{\lambda m}{2a}\right)^{1/2} \int_0^t \left[\frac{s\dot{\gamma}(s, \theta)}{\pi(t - t')}\right]^{1/2} dt' + \frac{1}{\sqrt{\pi t}} \int_0^\infty e^{-\xi^2/4t}\, \bar{\theta}_0(\xi)\, d\xi,$$

$$\tag{7.79}$$

where $\dot{\gamma}(s, \theta)$ is the constitutive function for the plastic strain rate, and $\bar{s}_0$, $\bar{\theta}_0$, and $\bar{v}_0$ are initial values for nondimensional stress, temperature, and velocity, respectively. The nondimensional constants m, λ, ρ, and a are determined

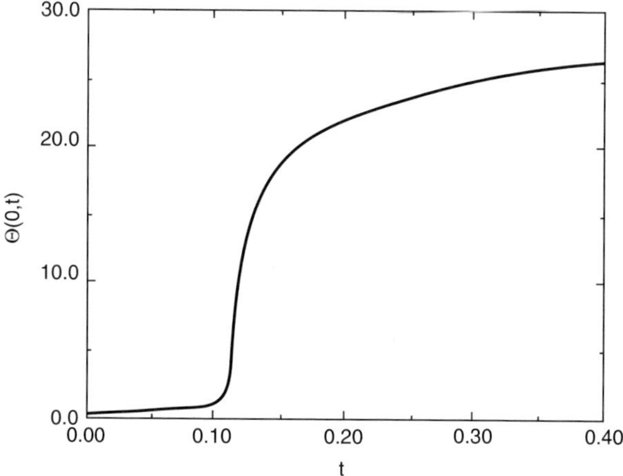

Figure 7.11. Typical development of temperature in the center of a shear band, according to the theory of DiLellio and Olmstead. Compare with the evolution shown in Figure 5.2; scales in the two figures are not the same, but clearly the general shapes are. (Reproduced from DiLellio and Olmstead 1997a, with permission from Elsevier Science.)

by the physical constants of the material and certain characteristic values of stress, strain rate, and temperature.

In nondimensional units, typical results for temperature and stress at the centerline are shown in Figures 7.11 and 7.12. In typical fashion they both

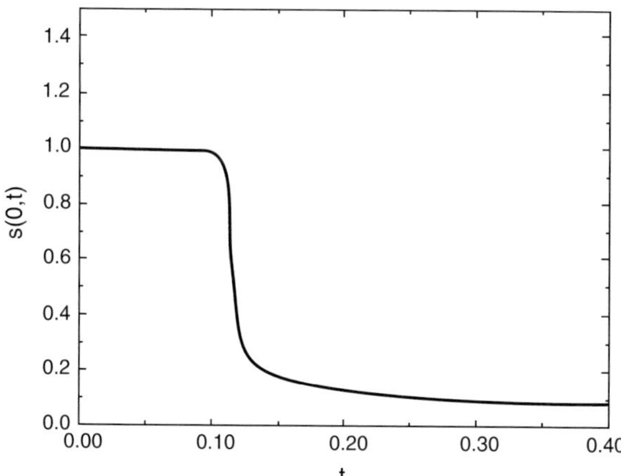

Figure 7.12. Typical development of stress in a shear band, according to the theory of DiLellio and Olmstead. Compare with the evolution shown in Figure 5.4; scales in the two figures are not the same, but clearly the general shapes are. (Reproduced from DiLellio and Olmstead 1997a, with permission from Elsevier Science.)

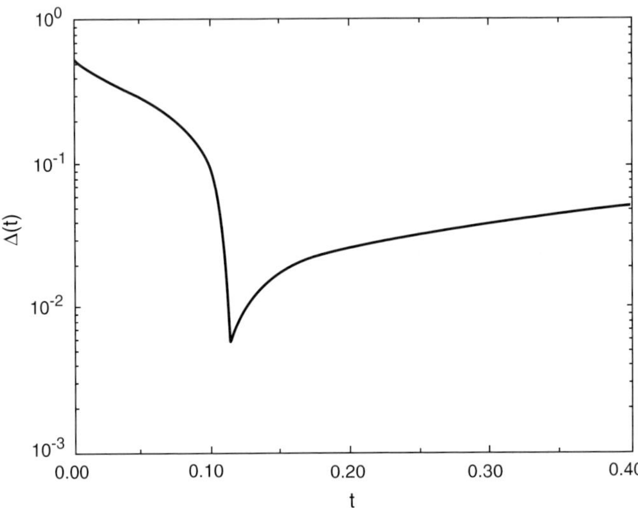

Figure 7.13. Typical development of the mechanical length scale of a shear band, according to the theory of DiLellio and Olmstead. Because exponential rather than linear softening was chosen, the width is a minimum just after localization and grows wider with continued straining. A similar result was reported by Walter (1992). (Reproduced from DiLellio and Olmstead 1997a, with permission from Elsevier Science.)

change slowly until rapid localization sets in. Then the temperature rises dramatically and the stress falls precipitously. Figure 7.13 shows how the characteristic halfwidth of the mechanical transition changes at the same time. These results show the same type of variation with time as may be obtained by a full finite-element calculation (e.g., see Walter 1992).

7.7 Concluding Remarks

As in Chapter 6, it again seems worthwhile to summarize the principal results of this chapter just in words. The intent of the chapter is to continue building a general picture of material instability and localization from examination of special cases. After the nonlinear equations are restated in nondimensional form, but with an addition to account for variations in the thickness of the sheared section, several exact and approximate solutions are given for a series of problems.

Adiabatic problems with vanishing thermal conductivity are considered first in Section 7.1. With a stress boundary condition given for a finite interval, a general and exact version of the solution presented by Molinari and Clifton

(1987) is found in implicit form by quadrature. Then for a velocity boundary condition given for a finite interval, after new independent and dependent variables are defined, an exact solution is found in implicit and parametric form. In this case the solution is expressed as a sequence of three quadratures (Wright 1990a, 1990b). Although the solution appears to be quite complex, it may easily be understood qualitatively by graphical means, and interpretions for three different kinds of perturbations are given. These results also show that for power law rate hardening the solution always becomes singular in the absence of heat conduction.

Section 7.2 discusses solutions for the velocity boundary condition when the material does not work harden but the thermal conductivity of the material does not vanish. First an exact solution is given for a simple, although rather degenerate case, demonstrating that strain softening by itself does not guarantee material instability and localization. Next an approximate solution is given for a material with linear thermal softening and power law rate hardening. The structure of the solution is similar in many ways to the case with no thermal conductivity, but now the solution requires solving a simple diffusion equation, a quadrature, and a nonautonomous ordinary differential equation. Again the result is in implicit and parametric form. This seems quite complicated, but further analysis leads to the conclusion that with proper nondimensionalization and linear scaling all nonwork hardening materials with a given strain rate sensitivity have virtually identical tendencies to localize, at least until the strain rate is high enough to introduce inertial effects. Furthermore, weak perturbations in strength, temperature, or web thickness are additive so that all have the exact same effect on localization when properly scaled. These predictions have been verified numerically, but they have not been verified by laboratory experiments.

An analysis of the structure of a fully formed band, first in a simple case and then in a general case, develops the idea that the structure of a shear band is canonical for all materials. One feature of the canonical structure is that distinct mechanical and thermal length scales are important in an adiabatic shear band. The mechanical scale gives a measure of the transition zone for the jump in velocity across the band. Therefore, it also is a measure for the width of the zone of high strain and strain rate. In contrast, the thermal scale gives a measure for the width of the zone of elevated temperature during band formation. Heat, which is generated in the zone of high strain rate, must be conducted away into the adjacent nondeforming and cooler region. This process is generally not nearly as rapid as the mechanical shearing, and therefore the thermal scale is much larger than the mechanical scale. Another scale, deduced directly from the equations rather than from solutions, was given by Dodd and Bai (1985). This third scale is shown to be the geometric mean of the other two. These matters are described in Sections 7.3, 7.4, and 7.5.

Finally a unified theory of band formation in the one-dimensional case, given by DiLellio and Olmstead (1997a, 1997b, 1998), is briefly described in Section 7.6. All of the main features that have been deduced from special cases also appear in their more general treatment, thus giving added confidence in the picture that has been contructed.

8

Two-dimensional experiments

In all of the experiments discussed in Chapter 1, a whole region of material is subjected to nearly uniform stress everywhere. As a consequence, when instability develops somewhere in the interior of the region, a complete adiabatic shear band forms essentially all at the same time. In most circumstances, however, a shear band actually forms first at a particular location and then propagates laterally with motion in the material much like a mode II or mode III crack. Unlike a crack, however, continuity of material is maintained across a shear band so that traction may still be transmitted from one side to the other at least for some finite amount of relative displacement between the two sides. This is essentially two-dimensional behavior because the adiabatic shear band grows by extending sidewise in its own plane. The experiments to be described in this chapter attempt to capture some of the features of two-dimensional growth and propagation while at the same time maintaining a certain measure of simplicity so that particular behaviors of shear bands may be isolated for closer study.

8.1 Kalthoff's Experiment

The first two-dimensional experiment with an adiabatic shear band, deliberately designed so that initiation and propagation could be observed, was devised by Kalthoff (1987) and Kalthoff and Winkler (1988). The experiment and much subsequent progress have been reviewed and summarized by Kalthoff (1990, 2000). A plate, containing one or two thin slits or fatigue cracks on one edge (simply called a precrack or crack from here on), is struck edge on by an impacting projectile immediately adjacent to the existing cracks. The experimental design for the case of two initial cracks is sketched in Figure 8.1. The impact drives the material into mode II motion so that as soon as the loading stress wave passes the end of the crack, there is a mode II stress concentration

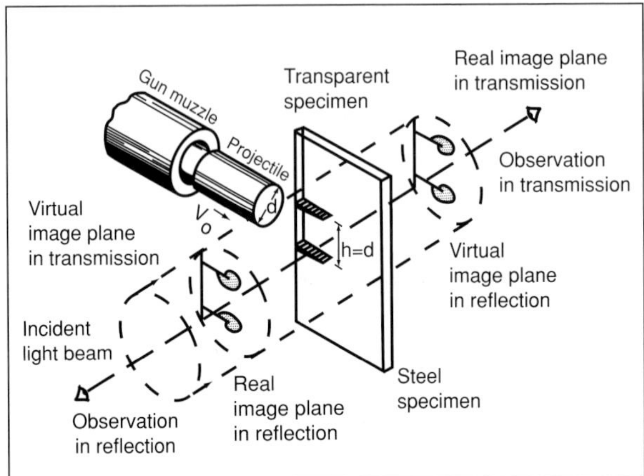

Figure 8.1. Schematic diagram of the Kalthoff experiment. A blunt projectile is launched from a gas gun so as to create an impact next to a precrack in the target plate. The deformation pattern at the tip of the propagating shear band or crack is observed optically, either in the reflection or transmission mode. (Reproduced from Kalthoff 2000, with permission from Kluwer Academic/Plenum Publishers).

at its tip. The subsequent behavior of the material is then observed by using high-speed photography and the shadow optical method of caustics. Stress intensity factors and speed of propagation may then be determined by analyzing the size and position of the shadow patterns in the sequence of photographs. The speed of the impacting projectile and the bluntness of the initial cracks may be varied, as well as the size of the target plate or specimen (see the references for further experimental details). Essentially three kinds of behavior have been observed: stable–brittle–ductile behavior with two transitional velocities of impact; brittle behavior under all achievable conditions; and ductile behavior under all conditions. The results for these three cases will be summarized in turn.

8.1.1 Stable–Brittle–Ductile Behavior

Experiments were performed on a high-strength maraging steel with a yield strength of approximately 2.1 GPa. For each fixed radius of curvature at the tip of the precrack it was found that there were three characteristic responses corresponding to three distinct ranges of impact velocity. At the lowest velocities the precrack was stable and did not extend. At intermediate velocities of impact a transition to brittle behavior occurred with a mode I crack departing laterally from the tip of the precrack. The angle of the new crack relative to the precrack was approximately 70° away from the impact side of the precrack, more or less in accordance with the direction of maximum tensile stress for a static mode II stress concentration in an elastic material. In every case the crack propagated all the way to the lateral boundary, and the failure surface had the typical rough surface and shear lips at the outer edges that is commonly associated with mode I cracks. Quasi-static, mode II loading on a preexisting crack with a different loading device also produced mode I cracks at $70°$ to the direction of loading.

At still higher velocities of impact a second transition occurred. This time the new failure surface propagated nearly straight ahead with only a small angle of deviation toward the compressive or impact side of the precrack. Upon examination the failure surface had a shiny, smeared over appearance. Furthermore, metallographs of sections taken perpendicular to the failure surface showed a region of increasing shear leading to a region of shearing separation between the two sides of the failure surface. Apparently an adiabatic shear band formed in the material and created a channel along which mode II failure could occur slightly later. In this regime the total length of the damage path increased with the speed of impact but except for one case did not extend to the far boundary. Once it is activated, the high-speed shearing mode of deformation appears to be far more efficient in dissipating the impact energy than is the mode I cracking in the intermediate regime.

When the experiments were repeated with differing radii at the tip of the preliminary crack, it was found that the two transition velocities separating the three types of deformation decreased as the radius decreased. For the high-speed regime it was also found that the length of the failure path increased as the radius of the crack tip decreased. Sketches of the failure paths are shown in Figure 8.2. The projectile struck just below the precrack. The dark lines between boxes show the approximate separation between the three categories of deformation. The upper left region is stable, the central region shows the mode I crack branching upward, and the lower right shows the shear band and failure surface moving ahead and slightly downward.

A more unified and precisely quantified picture of the various deformation regimes is found by plotting the length of the failure path versus a measure of the rate of loading at the notch tip, $v_0/\sqrt{\rho}$, where v_0 is the striking velocity and ρ is the radius of the notch tip. This is shown in Figure 8.3. Notice that mode I failure seems to lie in a well-defined interval below the region of shear failure. This fact argues for a well-defined switching of critical condition, as well.

8.1.2 Brittle Behavior

Similar impact experiments were also performed on epoxy specimens made of Araldite B with various radii of the initial notch tip and various impact

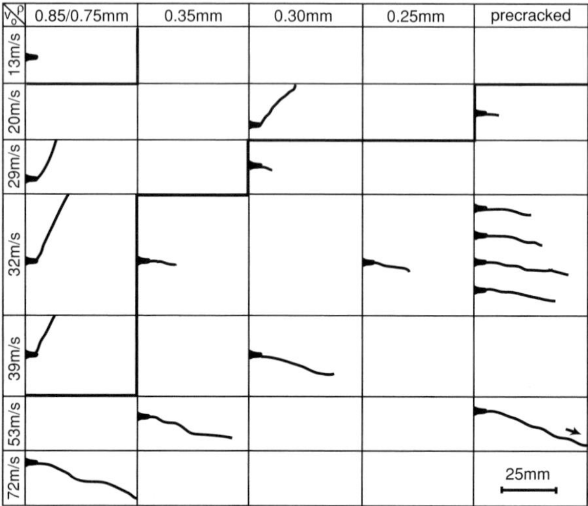

Figure 8.2. Damage paths observed in steel. The heavier lines show the approximate separation into the three regimes of stable, brittle, or shearing behavior. (Reproduced from Kalthoff 2000, with permission from Kluwer Academic/Plenum Publishers).

velocities. A lower limit on the impact velocity of approximately 9 m/s was imposed by the launch capabilities of the air gun. Similarly an upper limit of approximately 18 m/s was imposed by the complete failure and crushing of the specimen at the point of impact. All successfully tested specimens failed

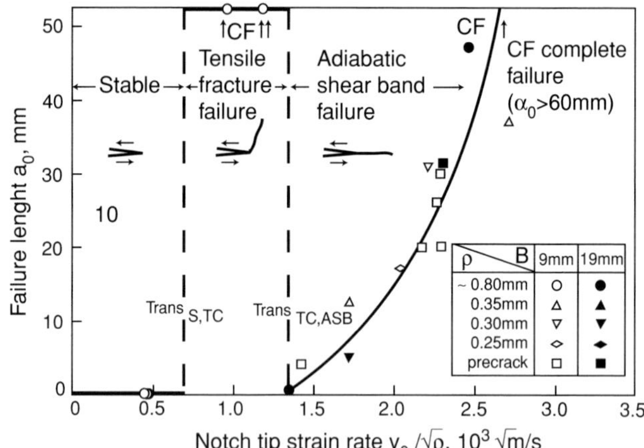

Figure 8.3. Failure lengths plotted against velocity of impact, rescaled by the square root of the radius of the initial precrack. (Reproduced from Kalthoff 2000, with permission from Kluwer Academic/Plenum Publishers).

by propagation of a mode I crack at approximately 70° to the orientation of the initial crack. This failure was exactly the same as that observed in the steel specimens at intermediate velocities of impact. Although transitional velocities were not observed, presumably there is one below which the precrack is stable, but if so, it falls outside the capability of the air gun.

8.1.3 Ductile Behavior

Edge impact experiments on 7075 Al alloy produced only mode II types of failures of limited length at all impact velocities from the lowest that could be obtained with the air gun up to approximately 70 m/s. This was also true for precracks with the sharpest tips, produced by fatigue. Quasi-static loading in mode II also produced only mode II failure. Even though mode I cracks can readily be produced with mode I loading conditions in this material, they never appeared when mode II loading was applied. Figure 8.4 summarizes the characteristic patterns of failure for all three materials under impact loading at various rates.

8.1.4 Other Results and Discussion

The experiments described above have been repeated by others for a variety of materials with more or less similar results. Mason, Rosakis, and Ravichandran

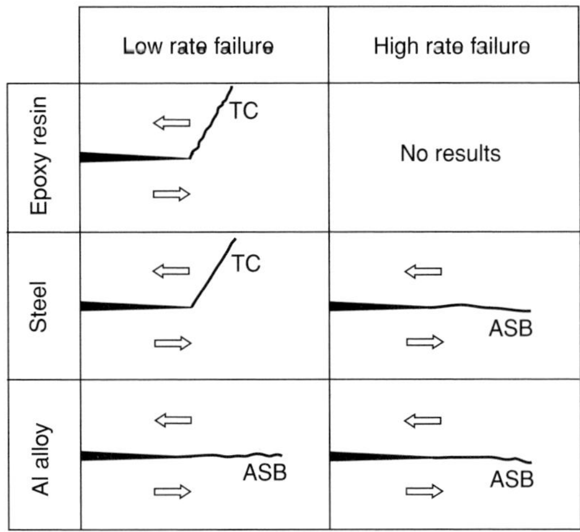

Figure 8.4. Modes of failure observed in three materials. (Reproduced from Kalthoff 2000, with permission from Kluwer Academic/Plenum Publishers).

(1994) performed impact experiments on C-300 steel and made full field measurements of the deformation at early times. After a delay of approximately 11 μs past the arrival of the compressive wave at the notch tip, a shear band appeared to grow straight ahead at the notch tip. The fringe pattern suggested that a mode II Dugdale plastic zone (identified as a shear band) formed at the tip of the crack when the stress intensity reached $K_{\mathrm{II}}^{d} = 140$ MPa$\sqrt{\mathrm{m}}$, that the band propagated at a speed of approximately 320 m/s, and that the mean stress on the shear band dropped from 1.6 Gpa at initiation to 1.3 Gpa before arresting, which is approximately a 20% reduction.

Zhou, Rosakis, and Ravichandran (1996) also performed impact experiments on C-300 steel and also on Ti-6Al-4V. Temperature measurements were made in the neighborhood of the propagating band. No mode I crack was observed to emanate from the tip of the precrack for any impact velocity. Instead the lowest transition velocity occurred for initiation of adiabatic shear bands, followed by shear cracking, that propagated nearly straight ahead of the precrack and slightly downward toward the impact side of the crack. The radius of the initial crack tip in these cases was 0.15 mm, which is smaller than any of the cases examined by Kalthoff and his colleagues for a steel similar to C-300. Presumably the smaller radius raised the effective loading rate, according to the scaling shown in Figure 8.3, so that mode I initiation was suppressed. After the shear bands arrested in C-300 steel, however, mode I cracks sometimes emanated from the point of arrest. Calculations by Zhou, Ravichandran, and Rosakis (1996) indicated that the late stage cracking was a consequence of reflected waves from the far boundaries. No late stage cracking was observed in the Ti alloy.

Experiments on polycarbonate by Ravi-Chandar (1995) showed nearly the same pattern as observed originally by Kalthoff and colleagues. With a slit width of 0.3 mm (halfwidth or radius of 0.15 mm) and a polycarbonate projectile as the striker, photoelastic experiments were performed over a range of impact velocities from approximately 20 m/s to about 55 m/s. No crack appeared for impacts below approximately 30 m/s, but a large inelastic zone appeared at the crack tip. For impacts in the range of 30–50 m/s a mode I crack departed from the tip of the precrack at an angle of approximately 66°. For impacts at velocities greater than 50 m/s a crack propagated straight ahead of the precrack for a distance of about 10 mm and then arrested. As in high strength steel, shear failure absorbed more energy than mode I failure.

Several computational studies have analyzed the Kalthoff experiment, such as those by Needleman and Tvergaard (1995), Zhou, Ravichandran, and Rosakis (1996), Batra and Gummalla (2000), and Batra and Ravinsankar (2000). The last is a three-dimensional analysis, and the others are two dimensional. The first paper includes void nucleation and growth through use of a modified Gurson model, but only the second one includes heat conduction. None of these analyses follows the process so as to observe actual localization, nor could they without heat conduction and much finer spatial resolution in their computations, but Zhou et al. attempt to simulate localization by introducing a critical strain, after which the shear band is represented as a Newtonian fluid along the line of anticipated failure. None of these analyses introduces actual shear failure and separation; however, all are successful in identifying regions around the notch tip where tensile stress or plastic shearing are particularly intense. Thus on the basis of the competition between a critical tensile stress and a critical equivalent plastic strain, it is possible to identify, at least qualitatively, the location and mechanism of subsequent failure. Furthermore, the likely path of the failure can also be identified.

As pointed out above, without heat conduction and adequate spatial resolution, it is impossible to simulate actual localization and propagation of an adiabatic shear band. The one-dimensional analog may be seen in Figure 5.3. In that figure as time and nominal strain increase, the strain rate increases also, but on a scale that is scarcely detectable compared with that in the fully formed band. Near a time that depends on the initial inhomogeneities, the strain rate is magnified by more than three orders of magnitude. Well prior to the transition to full localization, fivefold or tenfold amplifications of strain rate must occur, but that does not adequately signal the actual timing of full localization, which is far more severe.

To arrive at some notion of the situation in two dimensions, imagine the time axis in Figure 5.3 to be replaced by an axis in the direction of propagation. Then to a rough approximation the figure may be expected to represent the strain rate near the tip of a propagating shear band. Far before the arrival of full localization the strain rate will be amplified many times, and the local shear strain will become large. Of course, the real two-dimensional distribution would be significantly modified because of the extra spatial dimension. Nevertheless, from this point of view it would seem to be very difficult indeed to identify a "critical strain" at which localization occurs without further examination.

To date no calculations or analyses have been made that fully resolve the region immediately surrounding the tip of a propagating shear band, that include all relevant physics including heat conduction, and that make full connection to the driving forces and speed of propagation.

8.2 Thick-Walled Torsion Experiments

Chichili and Ramesh (1999) used a torsional Kolsky bar to drive an adiabatic shear band radially inward into a thick-walled tube in mode III motion. The bar

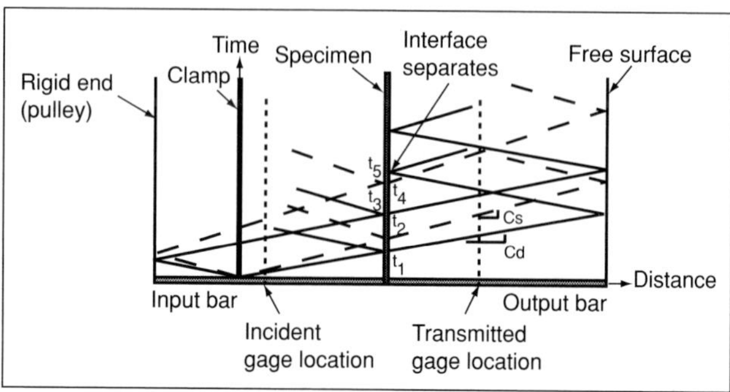

Figure 8.5. Wave diagram for the compression–torsion Kolsky bar in the recovery mode. (Reproduced from Chichilli and Ramesh 1999, with permission from ASME).

and experimental setup, as sketched in Figure 8.5, were designed so that both compression and torsion could be stored in one section of the input bar. When the clamp that stores the prestress is released, a combined compressive and torsional wave is sent through the specimen, which is held in place by a very thin layer of an epoxy adhesive. The design is such that the tensile reflection of the compressive wave breaks the glue at a predetermined time, thus simultaneously acting as a momentum trap for torsion. Once the bond between specimen and transmitter bar is broken, the specimen is subjected further to rigid motions only. In addition, only a single, well-defined torsional pulse, which is accompanied by a well-defined compressive pulse, has been used to load the specimen. Thus, the specimen may be recovered for metallurgical observation with a history that is well known and that may be correlated with microstructural changes. Figures 8.6 and 8.7 show typical compressive and torsional pulses.

The thick-walled specimen itself is shown in Figure 8.8. A small circumferential notch was placed in the center of the specimen so that the shear band would be located at the same axial location over its full length. Because of the design of the specimen, the deformation is not uniform either over its axial length or in the radial direction. Therefore, although the loading conditions on the ends of the specimen are well known from the experiment, the detailed distributions of stress and strain within the specimen are not known and must be determined by supplemental computations. The accuracy of the computations to a large extent depends on the accuracy of the constitutive equations for plastic flow, which must be known over a wide range of strain or work hardening, strain rate, and temperature. Thus, inhomogeneous experimentation and computation of this kind must be accompanied by an extensive program of testing under homogeneous conditions, unless the basic constitutive data are already known.

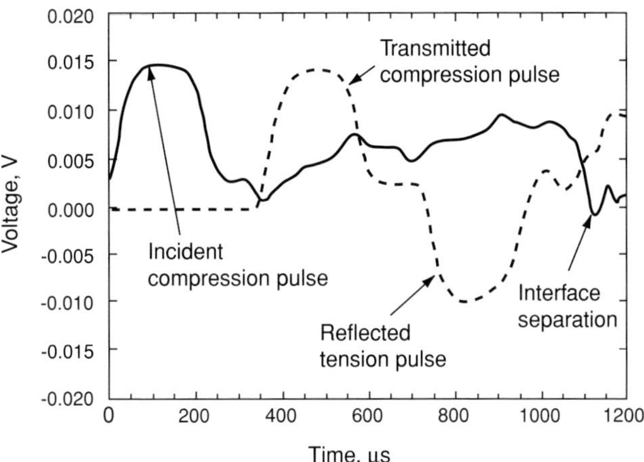

Figure 8.6. Compression signals for a test in the compression–torsion Kolsky bar. Note that the reflected tensile pulse causes the interface to separate from the specimen just after the shear wave has unloaded the specimen. (Reproduced from Chichilli and Ramesh 1999, with permission from ASME.)

In the original development of the thick-walled technique, shear banding in alpha titanium was studied. The necessary, supporting, constitutive work on alpha titanium was reported by Chichili, Ramesh, and Hemker (1998). This work covered a wide range of strain rates from quasi-static at $\mathcal{O}(10^{-5})$ s^{-1} to pressure–shear at $\mathcal{O}(10^{5})$ s^{-1}. Twinning, as well as crystallographic slip, was

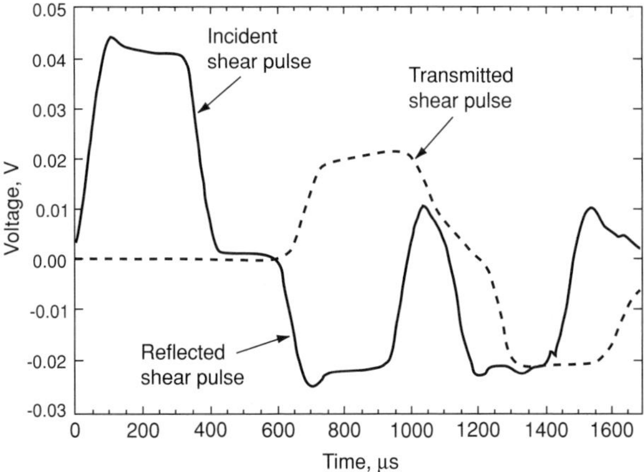

Figure 8.7. Shear signals with the specimen attached between the bars. The transmitted shear signal is trapped in the output bar, which separates from the specimen before the shear signal can return from the far end of the output bar. (Reproduced from Chichilli and Ramesh 1999, with permission from ASME.)

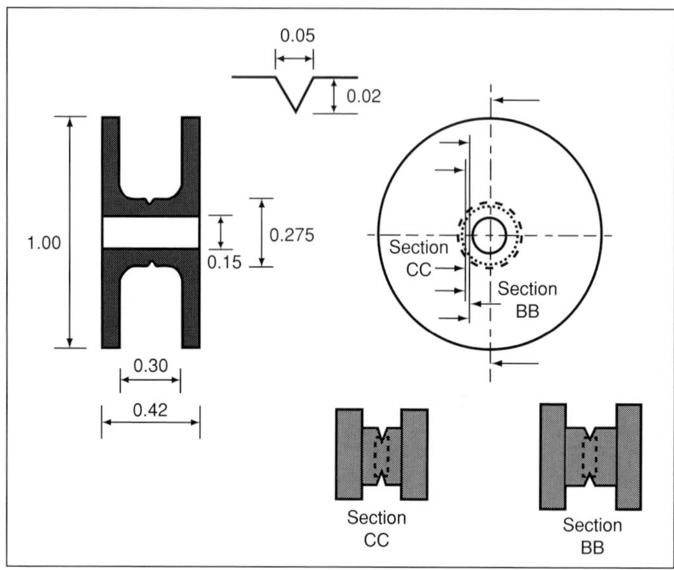

Figure 8.8. The thick-walled specimen used in the compression–torsion Kolsky bar has a small notch in the center of the gage section to initiate the shear band. Offset cross sections allow examination of the shear band at various depths from the notch. (Reproduced from Chichilli and Ramesh 1999, with permission from ASME.)

observed to play a strong role in the thermomechanical response of the material. It was found that monotonic loading could be captured fairly well with a Kocks–Mecking type of model, but that reload experiments at different temperatures and rates were less well modeled. Metallurgical and microstructural observations were reported by Chichili (1997).

In order to observe the effects of normal stress on the formation of shear bands, tests on the thick-walled specimens were made in several configurations that allowed varying combinations of a single torsional pulse and a single compressive pulse. With no axial compression, the specimen failed immediately upon formation of the shear band. With varying degrees of compression, however, failure could be wholly or partially suppressed by eliminating or delaying rapid growth of voids.

Metallurgical sections made parallel to the axis of the specimen, but offset as shown in Figure 8.8, showed the development of the shear band and voids at varying depths below the notch. With no axial compression there was evidence of extensive void formation, growth, and coalescence. With the addition of some axial compression, void formation and failure could be suppressed in varying degrees. On one hand, compression created the highest hydrostatic pressure just under the tip of the notch with lesser pressure toward the inner wall of the specimen. On the other hand, torsion created the largest shear strain and shear stress just under the notch with lesser amounts toward the inner wall. As a consequence, voids tended to form inside the specimen, rather than immediately under the notch. As might be expected, increasing compression at constant torque tended to decrease the number of voids, and increasing torque at constant compression tended to increase the number of voids.

Computations were made only for the adiabatic case and so were only partially successful. For example, although the region of high hydrostatic pressure was identified under the notch tip, the calculation predicted that the torque should drop as heating intensified and the band propagated inward, whereas in the experiment the torque continued to rise throughout. It might be speculated that heat conduction would delay both the loss of strength in the outer layers and the inward propagation of the band.

8.3 Collapse of a Thick-Walled Cylinder

A third type of experiment, designed to produce multiple interacting shear bands, was developed by Nesterenko and Bondar (1994) and applied to shear banding in Ti by Nesterenko, Meyers, and Wright (1998). The experimental setup is shown in Figure 8.9. The Ti specimen is surrounded by an inner and an outer Cu tube. The outer tube acts as a buffer between the explosive and the specimen, and the inner tube when fully collapsed acts to stop all further radial motion. The whole assembly of Ti and Cu tubes is surrounded by an explosive with a low velocity of detonation ($\sim$4000 m/s). The parts above and below the tubes and the point of initiation in the explosive are arranged so that the specimen will experience an axisymmetric impulse, which may be idealized as acting only in the negative radial direction.

Figure 8.10 schematically shows the expected behavior of the cylindrical Ti specimen, shown in cross section without the Cu tubes. As the cylinder first begins to collapse, it assumes a nearly axisymmetric motion, but as shear bands begin to form, axial symmetry is broken. The shear bands then continue to develop as the cylinder compresses further.

Initially the inmost Cu tube has an inner radius, R_i (in these experiments either 5.5 mm or 6.25 mm), the Ti has an inner radius of 7 mm, and the generic, initial, or material radius to an arbitrary interior point is denoted R. As the cylinder collapses, the inner and generic points map to $r_i(t)$ and $r(R, t)$, respectively. It will be assumed that all material is incompressible and behaves in

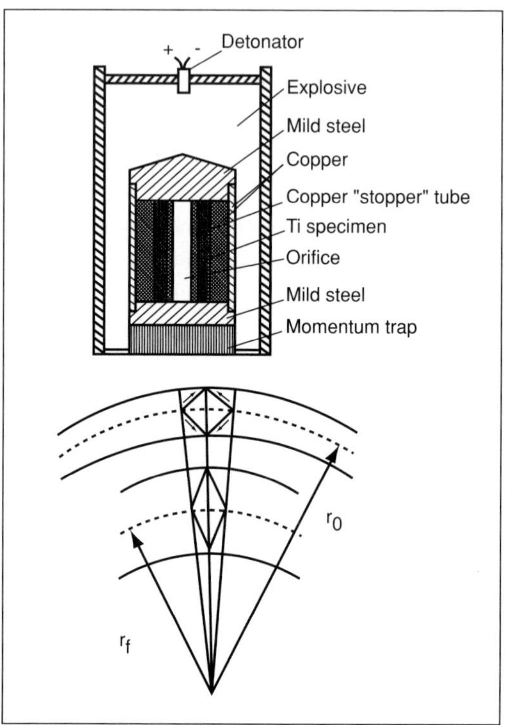

Figure 8.9. Experimental setup for the explosive collapse of a thick-walled cylinder. (Reproduced from Nesterenko et al. 1998, with permission from Elsevier Science.)

a rigid-plastic manner. Incompressibility in a two dimensional, axisymmetric motion may be expressed as

$$r^2 - r_i^2 = R^2 - R_i^2. \tag{8.1}$$

Consequently, the velocity gradient is determined from the kinematics to be a symmetric and deviatoric tensor:

$$\dot{\boldsymbol{F}}\boldsymbol{F}^{-1} = \boldsymbol{d} = \left\{ \begin{array}{ccc} \dfrac{\partial \dot{r}/\partial R}{\partial r/\partial R} & 0 & 0 \\ 0 & \dot{r}/r & 0 \\ 0 & 0 & 0 \end{array} \right\} = \left\{ \begin{array}{ccc} -r_i \dot{r}_i/r^2 & 0 & 0 \\ 0 & r_i \dot{r}_i/r^2 & 0 \\ 0 & 0 & 0 \end{array} \right\}. \tag{8.2}$$

The effective plastic shearing strain rate is

$$\dot{\gamma}_{\text{eff}}^{p} = \sqrt{2\,\text{tr}\,\boldsymbol{d}^2} = 2\left| \frac{r_i \dot{r}_i}{r^2} \right|. \tag{8.3}$$

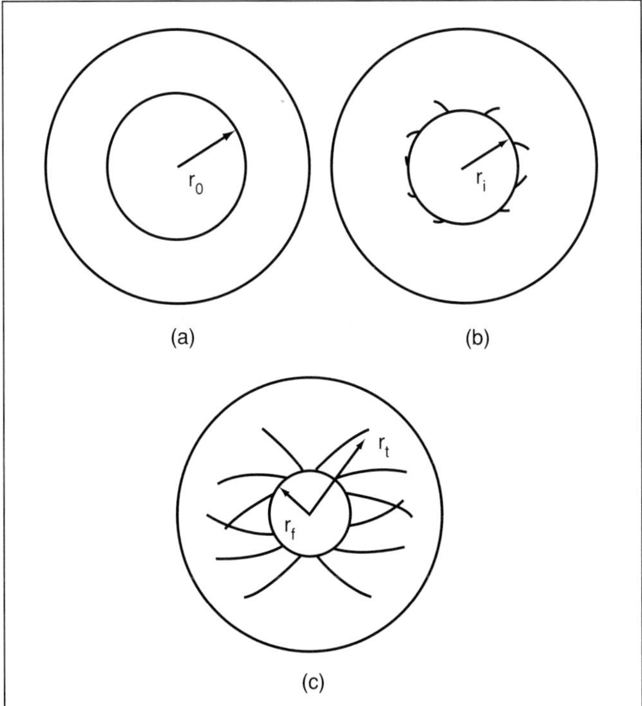

Figure 8.10. Schematic sequence of configurations as the cylinder collapses: panel a, initial configuration; panel b, shear bands initiate on the inner surface as the motion loses axial symmetry; and panel c, shear bands propagate outward along spiral trajectories as the cylinder collapses further. (Reproduced from Nesterenko et al. 1998, with permission from Elsevier Science.)

By application of Mohr's circle to the stretching tensor, the maximum shearing rate may be thought of as lying on planes at $\pm 45°$ to a radial line. The effective plastic shear strain at a material point is

$$\gamma_{\text{eff}}^p = 2 \int_0^t \left| \frac{r_i \dot{r}_i}{r_i^2 + R^2 - R_i^2} \right| dt = 2 \left| \ln \frac{r}{R} \right|. \tag{8.4}$$

In the paper by Nesterenko et al. (1998), a different scaling is used for effective plastic strain, namely $\varepsilon_{\text{eff}}^p = (2/\sqrt{3}) \ln(R/r)$. This sets the scaling to correspond to a simple tension test, rather than a pure shear test, as in (8.4).

Clearly from (8.3) the maximum shearing strain rate occurs on the inner radius, $\dot{\gamma}_{\text{eff}}^p = 2|\dot{r}_i/r_i|$, but the rate at any other material point is easily expressed. Suppose that when the inner radius of the Cu stopper tube collapses to zero (essentially), a material point in the Ti has radius r_f, where $r_f^2 = R_f^2 - R_i^2$.

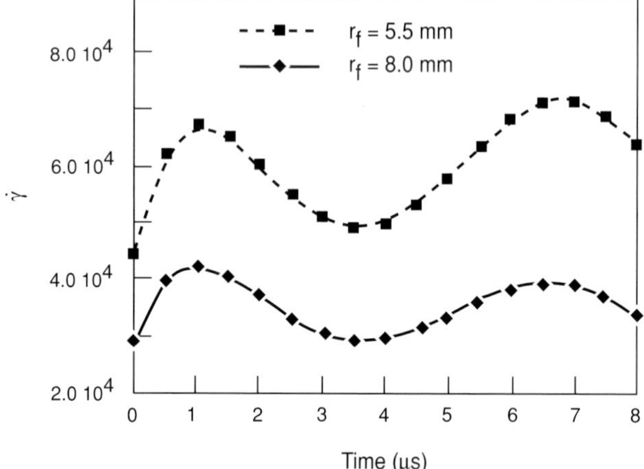

Figure 8.11. Shear strain rate for material that stops at two different final radii. (Reproduced from Nesterenko et al. 1998, with permission from Elsevier Science.)

Then after making use of r_f, rather than the corresponding initial coordinate R_f, to identify the material particle, one finds that the effective plastic strain rate at that particle may be expressed as

$$\dot{\gamma}_{\text{eff}}^{p}(R_f, t) = 2 \left| \frac{r_i \dot{r}_i}{r_i^2 + r_f^2} \right|, \quad r_i = R_i - \int_0^t |\dot{r}_i(t')| \, dt'. \tag{8.5}$$

Figure 8.11 shows the calculated strain rate histories for two typical material points, based on the interior velocity measurements of Nesterenko and Bondar (1994). They actually measured the radial velocity on the inner surface of a solid Cu thick-walled cylinder with the same overall configuration as the composite Cu-Ti-Cu cylinder shown in Figure 8.9. Although the replacement of some of the copper by the lighter Ti specimen will alter the interior velocity somewhat for the same impulsive loading, the velocity measurements on Cu alone may be used as a first approximation for the composite case.

Equation (8.5) holds throughout all the collapsing cylinders until shear bands begin to form at the inner radius of the Ti cylinder. Thereafter, the calculated strain rates will hold only in an average sense in the inner regions where the bands lie, although in the region beyond the shear bands, the strains and strain rates should still be nearly uniform. Final results for three specimens are shown in Figure 8.12. Note that the traces of fully developed shear bands spiral either clockwise or counterclockwise. At their tips they make angles of approximately $\pm 45°$ with a radial line, as expected, but they tend to align more closely with

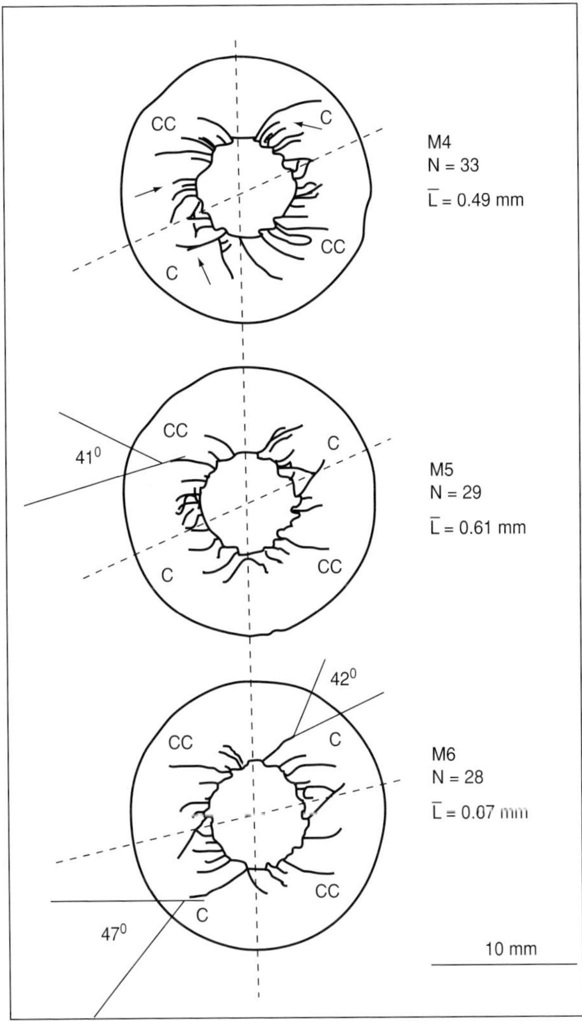

Figure 8.12. Final configuration of shear band traces for three different experiments. The initial inner radius of the Cu stopper tube was 6.25 mm. (Reproduced from Nesterenko et al. 1998, with permission from Elsevier Science.)

the radius itself at the inner surface. This also must be expected because fully formed shear bands lie on material surfaces, which are forced to rotate toward the radial direction as collapse continues.

The actual speed of propagation of the bands is unknown, but as in the case of mode III types of bands, as discussed in Section 8.2, they are forced toward an adverse region of lower strain and strain rate, so an extremely high speed

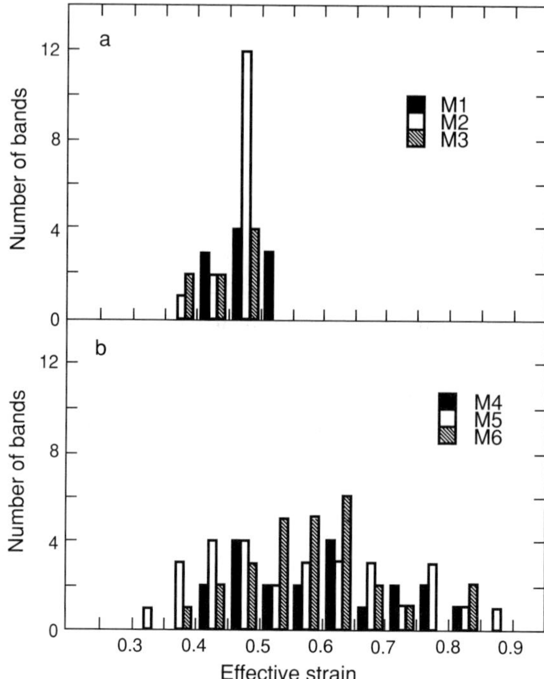

Figure 8.13. Distribution of effective strain at the tips of shear bands for two different inner radii of the Cu stopper tube: panel a, 5.5 mm; panel b, 6.25 mm. (Reproduced from Nesterenko et al. 1998, with permission from Elsevier Science.)

of propagation is not to be expected. A crude approximation may be obtained from Figures 8.11 and 8.12. Taking the length of the longest shear band to be approximately 4 mm and the total record to be approximately 8 μs long, one can estimate that the average speed of propagation is of the order of 500 m/s.

Figure 8.13 shows the distribution of shear bands at various effective strains, as calculated for the tips of the bands in the pattern frozen into the specimen by the two distinct sizes of stopper tubes. In the upper part of the figure the inner radius of the stopper tube was 5.5 mm, and in the lower part it was 6.25 mm. With the scaling used by Nesterenko et al. (1998), these produce a maximum effective strain in the Ti of approximately 0.55 and 0.92, respectively. The smallest strains actually correspond to points that are the farthest from the center of the cylinders and hence are the tips of the longest shear bands, whereas the largest strains correspond to bands that have nucleated just before the motion stopped. What the figure shows then is that new bands continue to nucleate as the motion continues, even as the bands that were earliest to nucleate continue to propagate. The upper part of the figure for the thicker stopper tube shows a large number of bands that nucleated just before motion stopped, but only a few bands that propagated out to the region of lowest strain. The lower part of the figure for the thinner stopper tube shows that the longest bands propagate to a region of approximately the same strain as do the longest bands for the thicker stopper tube, but that more bands continue to nucleate on the inner surface of the Ti right up to the time that motion stops. Furthermore, even though the two parts of the figure represent two different series of experiments, after comparing the numbers of bands at various strains in the upper and lower parts of the figure, one is tempted to conclude that not all of the bands that nucleate continue to propagate even though the mean strain near their tips must then increase as the inward motion continues. The longest bands do continue to propagate, but some others after their initial growth appear to have been shielded and to have stopped prematurely. Similarly it appears that after a certain number of bands have propagated outward, the regions between bands must be shielded to some extent to prevent large numbers of new bands from nucleating even though the average effective strain is very high.

In Chapter 6 formulas were given to predict the typical spacing that might be expected between shear bands. Equation (6.72), given by Grady and Kipp (1987), was based on the notion that shear localization would cause unloading in adjacent material, thus shielding it from formation of other shear bands. Equation (6.71), given by Wright and Ockendon (1996), was based on the idea that nucleating bands must be at least a minimum distance apart because of the influences of heat conduction and inertia. Although derived in very different ways, both formulas show similar scaling effects of thermal conductivity, heat capacity, thermal softening, material strength, and applied strain rate. They differ in numerical factors and in the fact that one includes strain rate sensitivity and the other does not. Molinari (1997) also developed a perturbation technique to estimate the spacing of shear bands. His method included work hardening, which the others do not include. When applied to Ti in the collapsing cylinders, the estimates of Grady and Kipp, Wright and Ockendon, Molinari, and the experimental result by Nesterenko et al. are

$$L_{\mathrm{GK}} = 3.3 \text{ mm}, \quad L_{\mathrm{WO}} = 0.52 \text{ mm}, \quad L_{\mathrm{MO}} = 0.75 \text{ mm}, \quad L_{\mathrm{EXP}} = 0.6 \text{ mm}$$

On one hand, all estimates are within an order of magnitude of the experimental result, which tends to give some confidence to the basic scalings. On the other hand, all three estimates are based on parallel bands, not on a cylindrical geometry. Furthermore, although the thermal conductivity and heat capacity are fairly well defined by handbook values, the strength, thermal softening, and strain rate sensitivity are less well defined. Furthermore, the strain rate and strain (hence, the radius and spacing) that prevail at band formation in the experiment are not very well defined either. Therefore, there must be considerable uncertainty, not only in the estimated values, but in the experimental one, as well.

Recently Xue, Nesterenko, and Meyers (2001) reported observing a distribution of shear bands with two distinct peaks in stainless steel that had been collapsed from the same geometry as described above. The longer bands had a spacing that seemed to correspond to the estimate of Grady and Kipp, but there was a second distribution of very short, recently nucleated bands with a spacing that seemed to correspond to the prediction of the perturbation techniques.

Experience to date, both analytical and experimental, seems to indicate that the nucleation of shear bands can be predicted, at least to some extent, by perturbation techniques. However, it also seems clear that the bands that initiate first may grow to such an extent that adjacent regions can be at least partially shielded as the stress in the band falls to lower levels. These and other multidimensional questions, such as speed and conditions for propagation, remain open. Progress to date on the two dimensional aspects of adiabatic shear bands is described in the final chapter. Three-dimensional aspects remain all but uninvestigated.

9

Two-dimensional solutions

The one-dimensional theory of adiabatic shear bands has been extensively described in previous chapters. A typical problem examines the evolution of a small perturbation in a finite domain and is usually meant to simulate a Kolsky bar test on a thin-walled tube with a short gage section. Adiabatic shear bands of finite extent often appear in bulk material as well, extending either to a boundary or ending rather diffusely in the interior of the sample. A band may appear as a thin, surface-like region of material, often with a poorly defined boundary, embedded in a three-dimensional continuum. The intersection of this surface with a metallographic section gives rise to the one dimensional, band-like structure that is observed experimentally.

In actual fact, a shear band is a dynamic entity, initiating at a point or in a small region of material and then spreading edgewise, somewhat like a crack. Even in the Kolsky bar experiment the band does not form simultaneously around the whole circumference of the specimen, as is tacitly assumed in the typical one-dimensional analysis. Rather, as has been observed by Marchand and Duffy (1988), it starts at a point (or possibly several points) and then spreads rapidly around the tube. Although the one-dimensional model is an effective way to represent the average response in a torsional Kolsky bar experiment, it cannot capture the crack-like propagation that is evident in other situations, such as any of the experiments described in the last chapter.

Shear bands become cracks if subjected to mode I opening motions; they only maintain their identity as shear bands with mode II or mode III sliding motions between the two sides of the band. Although numerical simulation has often been applied to the propagation of adiabatic shear bands, there have been relatively few analytical attempts to understand the essential features of propagation. The remainder of this chapter describes the few partly analytical solutions that have recently become available.

201

In one-dimensional problems it was found that temperature changes outside of the fully developed band are usually rather modest, and so the effect on the flow stress can be effectively modeled as linear in temperature. Inside the band a more complex thermal response might be appropriate. The multidimensional equations for a rigid-plastic material (i.e., $D^p = D$) with power law rate hardening may now be written in dimensional form as

$$\rho\dot{v} = \operatorname{div} S - \bar{\nabla}P, \qquad\qquad \text{momentum;}$$

$$\rho c\dot{\theta} = k\bar{\nabla}^2\theta + S{:}D, \qquad\qquad \text{energy;} \qquad\qquad (9.1)$$

$$S = \kappa_0(1 - a\theta)(bI)^m(2D)/I, \quad \text{flow law.}$$

Here S is the deviatoric stress tensor, P is the pressure, θ is the temperature difference from some arbitrary reference temperature, and D is the total stretching tensor whose second invariant is given by $I = \sqrt{2\operatorname{tr} D^2}$. The flow law has been written in a form that is an inversion of Equation (4.1). This will generally be possible for J_2-flow theories of viscoplasticity if the function Γ is monotonic in $\sqrt{J_2}$. The flow law in (9.1) is from Batra (1988) and is a generalization of the flow law introduced by Litonski (1977). The material will be assumed to be incompressible, $\operatorname{tr} D = 0$, and the pressure to be an arbitrary constant, $P \equiv \text{const.}$

9.1 Mode III: Antiplane Motion

Perhaps the first truly two-dimensional analysis was the study by Wright and Walter (1994, 1996) of a shear band in steady antiplane motion. Only the near-tip fields, exclusive of the tip itself, were analyzed by assuming a similarity solution of a particular form.

In antiplane motion only the out of plane component of velocity is nonzero, and all field variables are independent of the out of plane coordinate $\bar{z}$. Accordingly, the velocity and stretching components may be written as

$$v = (0, 0, w), \quad D = \left\{ \begin{array}{ccc} 0 & 0 & 1/2w_{\bar{x}} \\ 0 & 0 & 1/2w_{\bar{y}} \\ 1/2w_{\bar{x}} & 1/2w_{\bar{y}} & 0 \end{array} \right\}, \qquad (9.2)$$

and the deviatoric traction may be written as

$$Se_3 = s = se_w. \qquad\qquad (9.3)$$

Here, the unit vector $e_w \equiv \bar{\nabla}w/|\bar{\nabla}w|$ points in the direction of $\bar{\nabla}w$, s is the traction vector in the $\bar{x}$–$\bar{y}$ plane, and s is the magnitude of the traction. The balance equations and the flow law now reduce to the following coupled equations for w, θ, and s:

$$\rho w_{\bar{t}} = \bar{\nabla} \cdot (se_w), \quad \rho c\theta_{\bar{t}} = k\bar{\nabla}^2\theta + s|\bar{\nabla}w|, \quad s = \kappa_0 g(\theta)(b|\bar{\nabla}w|)^m. \quad (9.4)$$

In (9.4) the thermal softening factor $1 - a\theta$ has been replaced by $g(\theta) \equiv 1 - a\theta$.

Next let it be assumed that the band propagates edgewise at a constant speed, U, and that the motion as seen from the leading edge is steady. Thus, in the coordinate system

$$x = \bar{x} - U\bar{t}, \quad y = \bar{y}, \quad t = \bar{t} \tag{9.5}$$

the motion appears to be steady, and the equations of motion and flow law (with g rather than θ as a dependent variable) become

$$-\rho U w_x = \nabla \cdot (s e_w), \quad \rho c U g_x = -k\nabla^2 g + as|\nabla w|, \quad s = \kappa_0 g(\theta)(b|\nabla w|)^m. \tag{9.6}$$

9.1.1 Inertial Solution

To look for a solution that is valid near the tip of the band, it will be assumed that the dependent variables may be expressed as the product of the radius from the tip, $r = \sqrt{x^2 + y^2}$ raised to some power, times a function that depends only on the polar angle, $\phi = \tan^{-1}(y/x)$, measured from directly ahead of the band. It will also be assumed that heat conduction and the diffusion term $\nabla^2 g$ may be ignored except very close to the band itself, but inertia must be included. Thus, in an annulus that surrounds but does not include the tip of the band, that is, in $r_1 < r < r_2$ and $-\pi < \phi < +\pi$ for some small but fixed r_1 and r_2, the solution is assumed to have the form $w = Cr^\alpha W(\phi)$, and so on, where C and α are constants.

After the necessary exponents have been determined, a scaled similarity solution to (9.6) may be written as follows:

$$w = U \left\{ \frac{2m}{1+m} \frac{\rho c}{a\kappa_0} \right\}^{1/(1+m)} \left(\frac{r}{Ub} \right)^{m/(1+m)} W(\phi),$$

$$g = \frac{\rho U^2}{\kappa_0} \left\{ \frac{2m}{1+m} \frac{\rho c}{a\kappa_0} \right\}^{(1-m)/(1+m)} \left(\frac{r}{Ub} \right)^{2m/(1+m)} G(\phi),$$

$$\dot{\gamma} = \frac{1}{b} \left\{ \frac{2m}{1+m} \frac{\rho c}{a\kappa_0} \right\}^{1/(1+m)} \left(\frac{r}{Ub} \right)^{-1/(1+m)} \Gamma(\phi), \tag{9.7}$$

$$s = \rho U^2 \left\{ \frac{2m}{1+m} \frac{\rho c}{a\kappa_0} \right\}^{1/(1+m)} \left(\frac{r}{Ub} \right)^{m/(1+m)} S(\phi).$$

The functions W, G, Γ, and S are all nondimensional. The strain rate has been defined by $\dot{\gamma} = |\nabla w| = (w_r^2 + \frac{1}{r^2} w_\phi^2)^{1/2}$ and the stress from the third equation of (9.4). The leading scale factors are not arbitrary but have been chosen so that $S = G\Gamma^m$ and $\Gamma^2 = W_\phi^2 + \{[m/(1+m)]W\}^2$. Furthermore, the scaling also ensures a simplified form for the equations and initial conditions, as will become apparent below.

Although the radius in each equation of (9.7) is scaled by the distance Ub, which may be a length of the order of 10^6 m or more, in actual fact, the length scale is somewhat arbitrary because the radius could be multiplied inside the braces by an arbitrary constant without changing the value of the expression, provided that a proper compensating factor is used to multiply the whole right-hand side of each equation.

It is convenient to define an auxiliary dependent variable, Ψ, by $(1 + m)\Gamma \sin \Psi = mW$, or alternatively by $\Gamma \cos \Psi = W'$ where the prime indicates differentiation with respect to the argument ϕ. In physical terms Ψ is the angle between the unit vector e_w, defined from the gradient of w, and the unit transverse vector e_ϕ. Thus, $e_w \cdot e_x = \sin(\Psi - \phi)$ and $e_w \cdot e_y = \cos(\Psi - \phi)$.

After W and W' are eliminated in favor of Γ and Ψ, Equations (9.6) may be written as a system of first-order ordinary differential equations because r^α factors out of every equation.

$$(S \cos \Psi)' + \frac{1 + 2m}{1 + m} S \sin \Psi + \Gamma \sin(\Psi - \phi) = 0,$$

$$\sin \phi\, G' + \frac{2m}{1 + m}(\Gamma^{1+m} - \cos \phi)G = 0,$$

$$(\Gamma \sin \Psi)' - \frac{m}{1 + m}\Gamma \cos \Psi = 0, \tag{9.8}$$

$$S' - (G\Gamma^m)' = 0.$$

Again the prime indicates differentiation with respect to the angle ϕ.

Because the motion has been assumed to be antiplane, W is an odd function in ϕ, and therefore $W(0)=0$. Consequently from the definition of Ψ, its initial condition must be $\Psi(0) = 0$. The temperature and the magnitude of the stress must be even in ϕ and do not vanish at $\phi = 0$. The second equation of (9.8) reduces to $(\Gamma^{1+m} - 1)G = 0$ at $\phi = 0$, and therefore, if there is to be any solution at all, $\Gamma(0) = 1$. Only the initial values of stress and temperature are undetermined, so $S(0) = G(0) = G_0$ where the constant is arbitrary. In summary,

$$S(0) = G(0) = G_0, \quad \Psi(0) = 0, \quad \Gamma(0) = 1. \tag{9.9}$$

Equations (9.8) with initial conditions (9.9) constitute a universal set of equations that depend only on the strain rate sensitivity and the initial value for stress. It is a remarkable fact that all other physical parameters, including the speed of propagation, have been scaled out so that they appear only through the scale factors in (9.7).

The driving velocity, W, may be recovered from the definition for Ψ, and the transverse and parallel shear stresses are obtained from the decomposition of the traction, $s_{13} = s e_w \cdot e_x$ and $s_{23} = s e_w \cdot e_y$. For the nondimensional components we write

$$W = \frac{1+m}{m}\, \Gamma \sin \Psi, \quad S_{13} = S \sin(\Psi - \phi), \quad S_{23} = S \cos(\Psi - \phi). \quad (9.10)$$

Numerical solutions to Equations (9.8) were found by using standard methods. Typical results are shown in Figure 9.1 for the four primary dependent variables, W, G, Γ, and Ψ, as well as the two components of shear stress, S_{13} and S_{23}. In the figure, G_0 is held fixed at 1.0 while the strain rate sensitivity

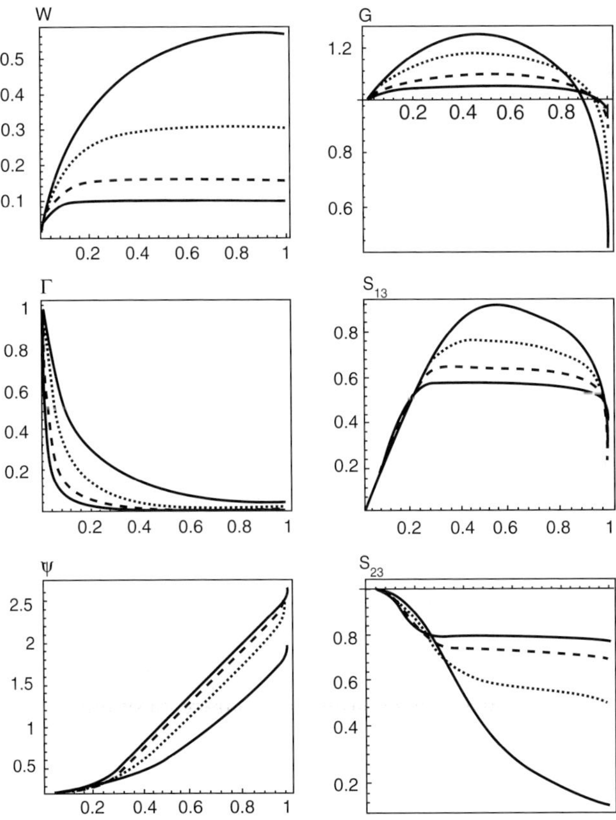

Figure 9.1. The effects of strain rate sensitivity on the characteristic nondimensional functions for antiplane motion of an adiabatic shear band are shown as functions of ϕ/π. Curves for values of $m = 0.01, 0.02, 0.05$, and 0.1 are shown, corresponding to the sequence of curves from bottom to top in the upper four figures, and top to bottom in the lower two. (Reproduced from Wright and Walter 1996, with permission from Elsevier Science.)

is varied. Besides a substantial effect on the level of the curves, increasing m also has a qualitative effect, generally smearing out the curves into more rounded shapes so that different regions are less clearly delineated from each other. The condition $G_0 = 1$ corresponds to fixing the reference temperature (where $\theta = 0$), the thermal sensitivity, a, and the reference strength, κ_0, to ambient conditions immediately ahead of the "tip" of the band.

As the angle, ϕ, increases, particle velocity rises rapidly from its initial value of zero ahead of the band. After some angle that increases with strain rate sensitivity, it takes on a nearly constant value, so as the strain rate sensitivity increases, the flat region becomes smaller, as seen in Figure 9.1.

Temperature, which can be understood qualitatively by flipping the G curve upside down since $\theta = (1 - g)/a$, shows a ridge of higher temperature ahead of the band with cooler regions to the sides. An indication of the shear band itself shows up as $\phi \to \pi$ where again the temperature rises rapidly, but in this region the similarity solution becomes invalid as heat conduction then has to be included. Increasing strain rate sensitivity strongly accentuates the leading ridge. The addition of heat conduction would be expected to have the effect of flattening and broadening the ridge, especially in the cases in which the ridge is more pronounced.

For the lower strain rate sensitivities, the driving shear stress, S_{23}, drops by 20% or so, but for the larger values, it drops as much as 80%. Here again it can easily be seen that the leading ridge is accentuated by increasing strain rate sensitivity.

Figures 9.2 and 9.3 show full solution surfaces, including the radial dependence, of the velocity, w, and of the driving shear stress, S_{23}, respectively. Here the rapid rise of velocity from zero ahead of the band to a plateau on the sides is readily visible, as is the leading wedge of higher stress. The scales in these two figures are arbitrary because compensating constants could be inserted into the length and amplitude scales of the complete formulas, as mentioned previously.

The speed of propagation of a shear band, which has been studied recently by numerical methods by Bonnet-Lebouvier, Molinari, and Lipinski (private communication), may also be found theoretically for the unconstrained case. Elimination of the radial dependence from the second and fourth equations of (9.7) yields the result $U = (s^+/\rho)\sqrt{(1 + m)/2m}\sqrt{a/g^+ c}$ where the superscript $+$ sign means that the quantity is to be evaluated immediately ahead of the band. The unit initial conditions have also been used. Now note that $a/g^+ = (-s_\theta/s)^+$, so that the expression for the speed of an unconstrained shear band may be written more generally as

$$U = \sqrt{\frac{s^+}{\rho}}\sqrt{\frac{-s_\theta^+}{\rho c}}\sqrt{\frac{1 + m}{2m}}. \tag{9.11}$$

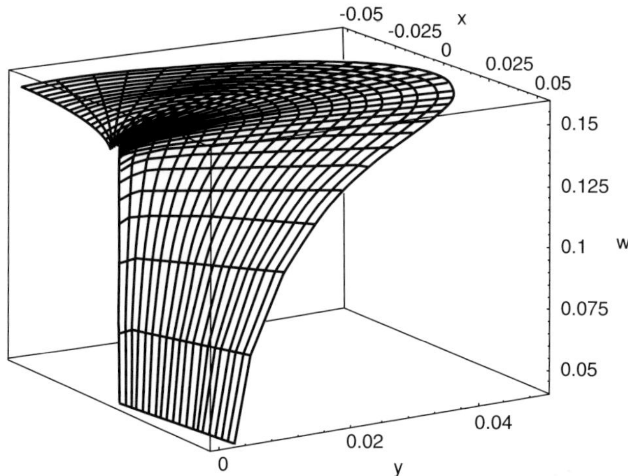

Figure 9.2. Typical solution surface for velocity, w, with arbitrary scales. The shear band extends to the left along the x axis from $x = 0$. (Reproduced from Wright and Walter 1996, with permission from Elsevier Science.)

The nondimensional term $-s_\theta^+/\rho c$ may be interpreted as the amount of softening of the flow stress per unit plastic work because the equation for the balance of energy in the adiabatic case may be written as $\rho c\, d\theta = s\, d\gamma^p$. Thus the nondimensional ratio of the softening index to the strain rate sensitivity is once again seen to be a dominant scaling parameter.

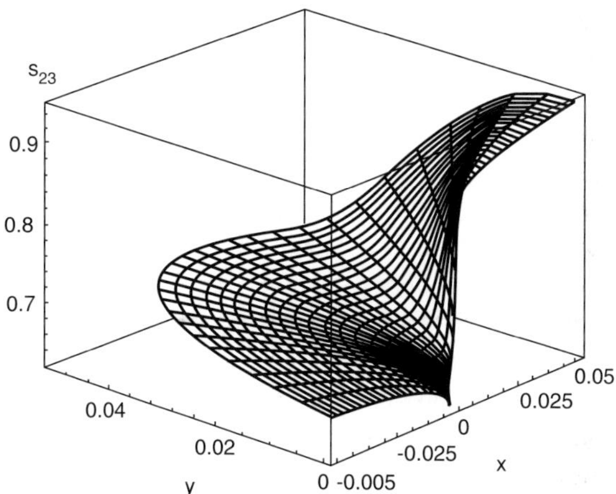

Figure 9.3. Typical solution surface for driving shear stress, s_{23}, with arbitrary scales. The shear band extends to the left along the x axis from $x = 0$. (Reproduced from Wright and Walter 1996, with permission from Elsevier Science.)

As a final remark, notice that once the constitutive behavior has been calibrated to the region directly in front of the shear band, there are no more arbitrary parameters still to be determined in the similarity solution (9.7). This occurs because the initial conditions for (9.8) then are all fixed, and the speed of propagation is determined by (9.11).

9.1.2 Core Solution

In the core of the shear band near $\phi = \pm\pi$, heat conduction more or less balances the heat source from plastic working and can no longer be ignored. In this region the limits of shear stress and velocity from the inertial solution in most cases should power the shear band in accordance with the canonical structure described in Chapter 7, Section 7.4. According to Equation (7.72), the strain rate and temperature in the center of the band, as well as the mechanical length scale, may all be expressed as functions of the applied shear traction and the jump in velocity across the band. For example, in the case of linear softening as in Equation (9.4),

$$g_c \equiv 1 - a\theta_c = \frac{\tau}{\kappa_0}\left(\frac{2km}{\kappa_0 ab}\right)^{m/(1-m)}\left(\frac{1}{2}\Delta v\right)^{-2m/(1-m)},$$

$$\lambda = \frac{a\tau}{2kmg_c}\left(\frac{1}{2}\Delta v\right)^2, \tag{9.12}$$

$$\delta = \frac{4kmg_c}{a\tau\,\Delta v},$$

where τ is the driving shear stress on the band. These equations were given by Wright and Walter (1996). Then from the asymptotic expression for temperature in (7.68) and the definition $\hat{a}_c = a/g_c$ from (7.56) and the flow law in (9.4), after a slight rearrangement, we have

$$g^{\text{core}} = g_c\{1 + 2m\ln[\cosh(y/\delta)]\}. \tag{9.13}$$

Because g_c and δ are known once the driving stress and jump in velocity are known, the whole structure of the shear band is known. In the present case, the driving shear stress is given by $\tau = s_{23}$ and half the jump in velocity is given by $1/2\Delta v = w$, where s_{23} and w are to be evaluated at $\phi = \pi^-$, that is to say, in the vicinity of the shear band. Figure 9.4 shows the core and nearby inertial solutions in a typical case. Of course, heat conduction will cause the inertial and core solutions to blend together as shown schematically by the dotted line in the figure.

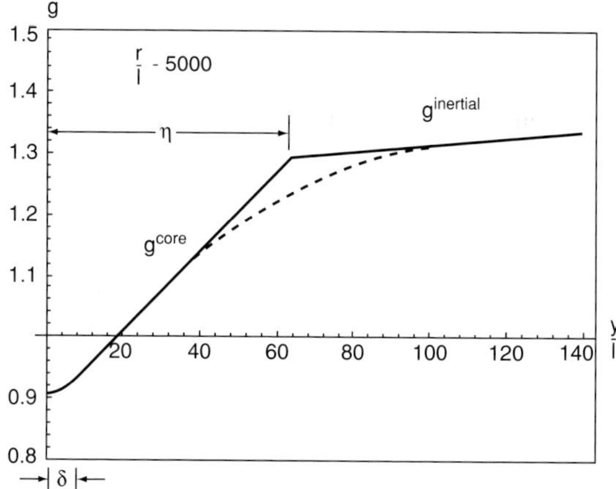

Figure 9.4. Typical thermal profile, showing the intersection of the inertial and core solutions. The dotted line sketches the effect of heat conduction in smoothing the two solutions. The mechanical and thermal widths are labeled δ and η, respectively. (Reproduced from Wright and Walter 1996, with permission from Elsevier Science.)

One measure of the thermal width of the shear band is the distance at which the temperature profiles from the two solutions intersect, as shown by η (rather than δ_{th}) in the figure. The points of intersection determine a curve in the x–y plane, which approximates the thermal trace of the shear band. In polar coordinates the curve is found from

$$g^{\text{core}}(r, \theta) = g^{\text{inertial}}(r, \theta). \tag{9.14}$$

First evaluate the argument in (9.13).

$$
\begin{aligned}
\frac{y}{\delta} &= \frac{r \sin \phi}{\delta} = r \sin \phi \frac{a(s_{23}/g_c)\,w}{2km} \\
&= \left\{ \frac{1}{1+m} \frac{r}{k/\rho cU} W^{1+m} \right\}^{1/(1-m)} \sin \phi,
\end{aligned}
\tag{9.15}
$$

where the ratio $s_{23}/g_c = \tau/g_c$ is taken from (9.12) and $1/2\Delta v = w$ is taken from (9.7). Note that the coefficient of thermal softening has disappeared from the formula and that, except for strain rate sensitivity, a natural length scale $\ell = k/\rho cU$ accounts for all other physical and kinematic properties that remain in the formula. In particular, strength, thermal softening, and the rate coefficient b have cancelled out of the formula exactly, and so do not appear anywhere.

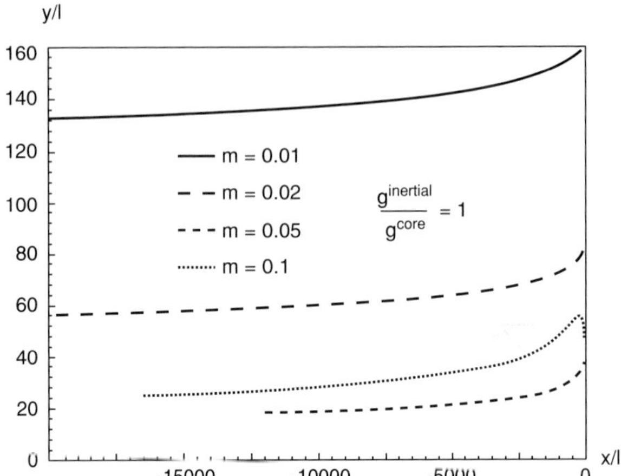

Figure 9.5. The effect of strain rate sensitivity, m, on the thermal width in units of length $\ell = k/\rho cU$. (Reproduced from Wright and Walter 1996, with permission from Elsevier Science.)

Now after taking g^{core} from (9.13), g_c from (9.12), and g^{inertial}, w, and s_{12} from (9.7), Equation (9.14) reduces to

$$1 = \left(\frac{1+m}{r/\ell}\right)^{m/(1-m)} \frac{S_{23}}{GW^{2m/(1-m)}}$$

$$\times \left\{1 + 2m\ln\cosh\left[\left(\frac{r/\ell}{1+m}W^{1+m}\right)^{1/(1-m)}\sin\phi\right]\right\}. \qquad (9.16)$$

Again the length scale ℓ has appeared in a natural manner. Because S_{23}, G, and W depend only on ϕ, Equation (9.16) is an implicit equation for the curve, $r(\phi)$, that defines the thermal width of the shear band.

Solutions of (9.16) are shown in Figure 9.5, width $\delta_{\text{th}}(x/\ell)$ being plotted in x–y coordinates in units of length, $\ell = k/\rho cU$. For steel with $k \approx 50$ W/m C$^\circ$, $\rho \approx 8 \times 10^3$ kg/m^3, $c \approx 500$ J/kg C$^\circ$, and $U \approx 250$ m/s, the length scale works out to be $\ell = 50$ nm. The figure shows that a shear band is indeed a long narrow structure that does not vary much in width for as much as a millimeter ($x/\ell = -20,000$ in the present case) behind the leading edge. Furthermore, the width is strongly affected by the strain rate sensitivity, but it can be shown that it is little affected by the leading temperature. As indicated by the natural length scale ℓ, the actual dimensional width will also be affected by material properties in the ratio $k/\rho c$, as well as by the speed of propagation U.

The thermal length scale δ_{th}, defined pictorially in Figure 9.4, may also be estimated as $\delta_{\text{th}} = |\Delta\theta/\theta_y|$, where $\Delta\theta$ is the temperature difference between the center of the band and the adjacent ambient region, and θ_y is the asymptotic slope. As discussed in Chapter 7, Section 7.5, the estimate for δ_{th} is $(\hat{a}_c \Delta\theta/2m)\delta_{\text{mech}}$ and is larger than either the mechanical or the Dodd–Bai length scale, $\delta_{\text{mech}} \leq \delta_{\text{DB}} \leq \delta_{\text{th}}$.

In computations, if it is desired to resolve the shear band numerically, it is clear that the grid must be finer than the smallest length scale, which often is of the order of 1 μm, but which in any case can be estimated whatever the problem at hand. The authors of much computational work that has been reported in the literature seem to be unaware of the small size of the mechanical scale that must be resolved if complete fidelity is to be obtained. In one-dimensional problems the penalty paid for insufficient resolution is an apparent strength of the material that is only an artifact of the gridding. The length scale of the computational grid in effect becomes an internal length scale of the material. Presumably an insufficiently refined grid has the same effect in two dimensions, as well, but to the author's knowledge there have been no studies on this point.

9.2 Mode II: In-Plane Motion

Chen and Batra (1999) performed a similar, but far more complex, analysis of a shear band propagating in the direction of the dominant shearing for a rigid, viscoplastic material with linear thermal softening. Equations (9.1) are again applicable, but now the motion is assumed to lie in the plane of propagation. In polar coordinates $(\bar{r}, \bar{\phi})$ the velocity, projected onto unit vectors in the radial and transverse directions, may be written $v = u e_{\bar{r}} + v e_{\bar{\phi}}$. The acceleration $\dot{v}$ in expanded form is

$$\dot{v} = \frac{\partial v}{\partial \bar{t}} + v \cdot \bar{\nabla} v = \dot{u} e_{\bar{r}} + \dot{v} e_{\bar{\phi}},$$

$$= \left(\frac{\partial u}{\partial \bar{t}} + v \cdot \bar{\nabla} u \right) e_{\bar{r}} + \left(\frac{\partial v}{\partial \bar{t}} + v \cdot \bar{\nabla} v \right) e_{\bar{\phi}}. \tag{9.17}$$

The extra complexity arises because now there are two independent components of motion and two components of the momentum equation, instead of only one of each as in the previous section. Some of the extra complexity may be forestalled by the use of a stream function, χ, so that incompressibility, $\bar{\nabla} \cdot v = 0$, is automatically enforced.

$$v = \bar{\nabla} \times (\chi e_z) = \frac{1}{\bar{r}} \frac{\partial \chi}{\partial \bar{\phi}} e_{\bar{r}} - \frac{\partial \chi}{\partial \bar{r}} e_{\bar{\phi}}. \tag{9.18}$$

As before it will be assumed that the motion appears steady to an observer who translates with the tip of the shear band at a constant velocity, U. A moving polar coordinate system attached to the tip is related to the original coordinates by

$$r = [(\bar{r}\cos\bar{\phi} - U\bar{t})^2 + (\bar{r}\sin\bar{\phi})^2]^{1/2},$$

$$\phi = \tan^{-1}\left(\frac{\bar{r}\sin\bar{\phi}}{\bar{r}\cos\bar{\phi} - U\bar{t}}\right), \tag{9.19}$$

$$t = \bar{t}.$$

At time $\bar{t} = 0$ the two coordinate systems coincide. The components of particle velocity resolved along the two sets of axes also coincide at that moment. As a consequence the two spatial derivatives also coincide initially, $(\partial/\partial\bar{r}) = (\partial/\partial r)$ and $(\partial/\partial\bar{\phi}) = (\partial/\partial\phi)$, but the temporal derivative becomes $(\partial/\partial\bar{t}) = (\partial/\partial t) - U\cos\phi(\partial/\partial r) + (U\sin\phi/r)(\partial/\partial\phi)$. Because it has been assumed that the fields appear to be steady, the time derivative, $\partial/\partial t$, may be dropped. Thus the material time derivatives that appear on the left-hand sides of (9.1) for the steady, translating case may be written, for example, as

$$\dot{u} = (u - U\cos\phi)\frac{\partial u}{\partial r} + (v + U\sin\phi)\frac{1}{r}\frac{\partial u}{\partial\phi}, \tag{9.20}$$

with similar expressions for $\dot{v}$ and $\dot{g}$ (or $\dot{\theta}$). This expression is more complicated than the acceleration term found for the case of mode III shearing in (9.6) because of the presence here of the quadratic convective terms, which are absent in mode III. However, the shear band tip can move very rapidly, at many tens or even a few hundreds of meters per second, even though the particle velocity is rather small, perhaps only a few meters per second. At any rate it seems reasonable to assume that $U \gg (u^2 + v^2)^{1/2}$ so that a good approximation for the acceleration is

$$\dot{u} = -U\cos\phi\frac{\partial u}{\partial r} + \frac{U\sin\phi}{r}\frac{\partial v}{\partial\phi}. \tag{9.21}$$

The procedure is now exactly the same as in the case of mode III motion. In the inertial region (a ring that surrounds but does not include the tip, $r_1 \leq r \leq r_2$, for some small but finite r_1, r_2, and $-\pi < \phi < +\pi$) the heat conduction term is neglected, and it is assumed that all fields have similarity solutions that are proportional to r raised to some power times a function of angle only. For example, the two fundamental quantities χ and $g = 1 - a\theta$ may be expressed as

$$\chi = bU^2\left(\frac{\rho c}{a\kappa_0}\right)^{1/(1+m)}\left(\frac{r}{Ub}\right)^{(1+2m)/(1+m)}X(\phi),$$

$$g = \frac{\rho U^2}{\kappa_0}\left(\frac{\rho c}{a\kappa_0}\right)^{(1-m)/(1+m)}\left(\frac{r}{Ub}\right)^{2m/(1+m)}G(\phi). \tag{9.22}$$

All other quantities may be developed from these two. Some of the more important quantities are

$$u = \frac{1}{r}\frac{\partial \chi}{\partial \phi} = U \left(\frac{\rho c}{a\kappa_0}\right)^{1/(1+m)} \left(\frac{r}{Ub}\right)^{m/(1+m)} X'(\phi),$$

$$v = -\frac{\partial \chi}{\partial r} = -\frac{1+2m}{1+m} U \left(\frac{\rho c}{a\kappa_0}\right)^{1/(1+m)} \left(\frac{r}{Ub}\right)^{m/(1+m)} X(\phi),$$

$$I = 2(\operatorname{tr} \mathbf{D}^2)^{1/2} = \frac{1}{b}\left(\frac{\rho c}{a\kappa_0}\right)^{1/(1+m)} \left(\frac{r}{Ub}\right)^{-1/(1+m)} \Gamma(\phi),$$

$$S = (\operatorname{tr} \mathbf{S}^2)^{1/2} = \kappa_0 g(bI)^m = \rho U^2 \left(\frac{\rho c}{a\kappa_0}\right)^{1/(1+m)} \left(\frac{r}{Ub}\right)^{m/(1+m)} G(\phi)\Gamma^m(\phi),$$

$$(9.23)$$

where the nondimensional invariant of the stretching tensor, Γ, is given by

$$\Gamma(\phi) = \left\{\left[\frac{2mX'(\phi)}{1+m}\right]^2 + \left[X''(\phi) + \frac{1+2m}{(1+m)^2}X(\phi)\right]^2\right\}^{1/2}. \qquad (9.24)$$

Compare the scaling in Equations (9.22) and (9.23) with the scaling in Equations (9.7). The powers of r and of the nondimensional number $(\rho c/a\kappa_0)$ are identical for comparable quantities in both the mode II and mode III cases, but the expression $[2m/(1+m)]$ raised to the approximate power is missing. This last fact affects the initial values at $\phi = 0$.

When these expressions are substituted into the equations of motion, all powers of r cancel out, as do the leading coefficients, leaving only a set of coupled ordinary differential equations for $G(\phi)$ and $X(\phi)$. Chen and Batra write these equations as

$$(G\Gamma^{m-1}X_1)'' + 4\frac{m+2m^2}{(1+m)^2}(G\Gamma^{m-1}X')' - \frac{2m+3m^2}{(1+m)^2}(G\Gamma^{m-1}X_1)$$

$$= X_2'' + \left(\frac{m}{1+m}\right)^2 X_2, \quad G\Gamma^{1+m} = \frac{2m}{1+m}\cos\phi G - \sin\phi G', \qquad (9.25)$$

where the auxiliary functions $X_1(\phi)$ and $X_2(\phi)$ are given by

$$X_1(\phi) = X''(\phi) + \frac{1+2m}{(1+m)^2}X(\phi),$$

$$X_2(\phi) = \sin\phi X'(\phi) - \frac{1+2m}{1+m}\cos\phi X(\phi). \qquad (9.26)$$

Again, as in the case of antiplane shear, all physical constants except m drop out of the equations so that they have a kind of universality for different materials.

Boundary and initial conditions are suggested by examining the jump conditions across the shear band, treated as a singular surface embedded in the material. Because a shear band is a material surface, its velocity in the direction normal to the band is the same as the normal component of the material velocity, calculated on either side of the band; that is, $\boldsymbol{n} \cdot \boldsymbol{v}^+ = \boldsymbol{n} \cdot \boldsymbol{v}^- = v_n$. As a consequence, the jump conditions that express the balance laws reduce to

$$[\boldsymbol{v} \cdot \boldsymbol{n}] = 0,$$
$$[\boldsymbol{Tn}] = 0, \qquad (9.27)$$
$$[\boldsymbol{q} \cdot \boldsymbol{n}] + (\boldsymbol{Sn}) \cdot [\boldsymbol{v}] = 0.$$

where the square brackets indicate the difference of the enclosed quantity from one side of the band to the other. The first equation of (9.27) states that the tangential components of velocity may be discontinuous across a shear band, but not the normal component; the second states that the tractions on the plane of the shear band must be continuous; and the third states that the rate of work done by the deviatoric shearing tractions against the jump in the velocity must be balanced by the jump in the heat flux that carries heat away from the surface.

The first equation of (9.27) is satisfied if the reduced stream function $X(\phi)$ is an even function, making the radial component of velocity odd, and the transverse component even in the angle ϕ. The shearing component of the second equation of (9.27) will be satisfied if $G(\phi)$ is also an even function, making both the shear stress and the temperature even in ϕ. These two requirements are also sufficient to satisfy the third equation of (9.27), but because the solution being constructed does not allow for heat conduction, this third condition is irrelevant for the moment anyway. It becomes relevant when the core of the shear band is constructed as a boundary layer. Then the condition will be satisfied by the outer limits of the boundary layer solution.

The last remaining jump condition to be satisfied is the continuity of the normal stress $T_{\phi\phi}$ across the shear band. However, because $T_{\phi\phi}$ is odd in ϕ, the only possibility is that it must vanish at $\phi = \pm\pi$. In summary, the conditions deduced so far from the jump conditions are

$$X'(0) = X'''(0) = G'(0) = 0, \quad T_{\phi\phi}(\pm\phi) = 0. \qquad (9.28)$$

Equations (9.25) are fourth order in X and second order in G, so two more boundary conditions are still needed. As in the antiplane case, one of the remaining conditions is found by choosing $\Gamma(0)$ in such a way that $G_0 \neq 0$ is a permissible choice. At $\phi = 0$ the second equation of (9.25) reduces to $G(0)[\Gamma^{1+m}(0) - 2m/(1+m)] = 0$. If $G(0)$ is to be finite at $\phi = 0$, the term in

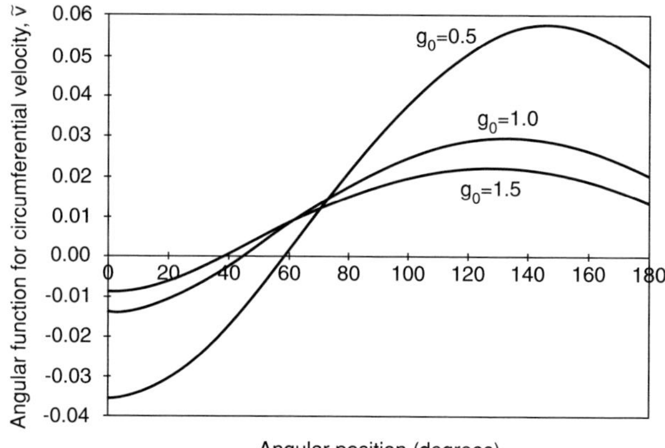

Figure 9.6. Circumferential velocity. Note the negative value at $\phi = 0$, which indicates that the path of the shear band must deviate from straight ahead. (Reproduced from Chen and Batra 1999, with permission from Elsevier Science.)

square brackets must vanish, which gives for the fifth boundary condition

$$X''(0) = \left(\frac{2m}{1+m}\right)^{1/(1+m)} - \frac{1+2m}{(1+m)^2}X(0) \tag{9.29}$$

after (9.24) has been used to express $\Gamma(0)$. The last condition is specified when $G(0) = G_0$ is chosen as a free parameter. In a complete problem it would have to be chosen so as to be consistent with the external flow. The actual method of solution of (9.25) used by Chen and Batra was to cast the problem as an initial value problem and then to vary $X(0)$ until the last of (9.28) was satisfied. The speed of propagation may be calculated in the same manner as before with the exact same results when the difference in scaling between (9.23) and (9.7) is taken into account.

Typical results are shown in Figures 9.6–9.10. Figures 9.6 and 9.7 show the radial and transverse components of velocity. The transverse component, shown in 9.6, is an even function of the polar angle so the normal velocity at the shear band is automatically continuous. The fact that the transverse velocity is slightly negative at $\phi = 0$ shows that the shear band has a tendency to veer away from its nominal direction of advance and toward the side with the driving shear and compressive stresses. The radial velocity, shown in Figure 9.7, is odd in the polar angle and therefore has a jump at $\phi = \pi$. It is this jump in velocity that, together with the shear traction, drives the shear band in accordance with the canonical structure, as explained in the Chapter 7, Section 7.4.

Figure 9.8 shows the effective strain rate around the tip of the advancing shear band. The rate is highest in a leading wedge ahead of the band, and the effective opening angle of the wedge increases with increasing strain rate sensitivity.

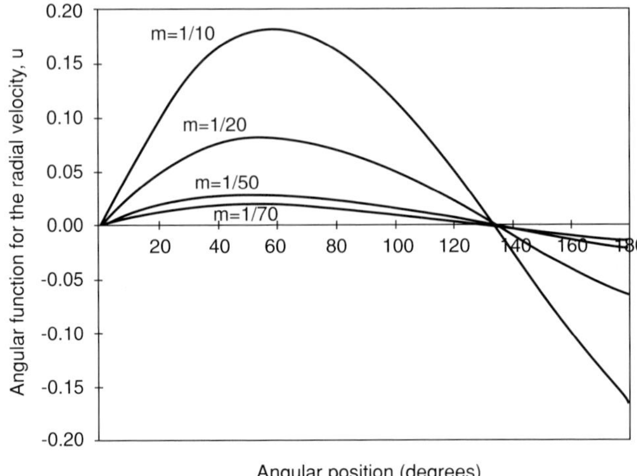

Figure 9.7. Radial velocity for different strain rate sensitivities. At $\phi = 0$ the radial velocity vanishes, but at $\phi = \pi$ its value is finite, which indicates that there must be a jump in the tangential velocity across the shear band. (Reproduced from Chen and Batra 1999, with permission from Elsevier Science.)

Figure 9.9 shows the effective stress and the three polar components of stress. The polar shear stress is even in the polar angle and therefore is continuous at $\phi = \pi$, as it should be so that the shear traction is continuous there. The transverse and radial stresses are odd in the polar angle so both vanish at $\phi = 0$. The transverse or hoop stress is made to vanish at $\phi = \pi$ so that the normal traction is continuous there, but there is no such requirement on the radial stress, which has a substantial jump across the shear band. On one side, there

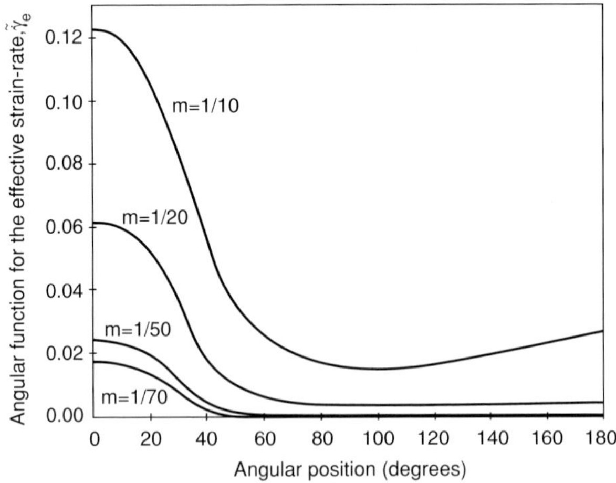

Figure 9.8. Effective plastic strain rate for different strain rate sensitivities. (Reproduced from Chen and Batra 1999, with permission from Elsevier Science.)

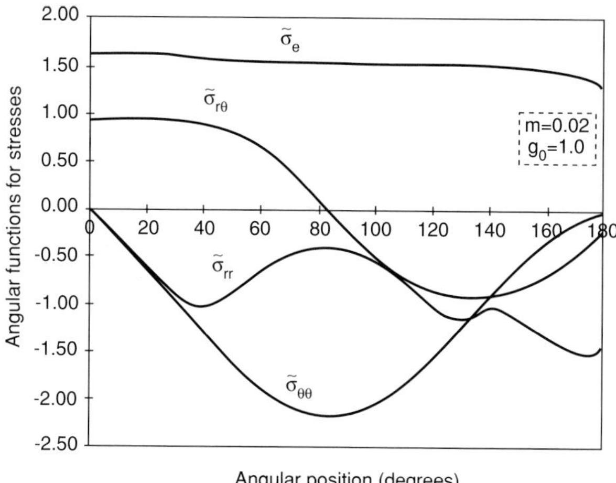

Figure 9.9. Angular distribution of stress components. Only shear stress is nonzero at $\phi = 0$. (Reproduced from Chen and Batra 1999, with permission from Elsevier Science.)

is a driving compression stress, and on the other side there is a tensile stress of equal magnitude.

Finally, Figure 9.10 shows the leading wedge of higher temperature (lower g) that precedes the shear band near $\phi = 0$. The wedge is more pronounced for higher strain rate sensitivity in agreement with the wedge of elevated strain rate.

The overall mode II behavior is similar to that of a mode III shear band in every aspect, besides the basic scaling as previously noted concerning Equations (9.22) and (9.23). Note especially the leading wedge of strain rate

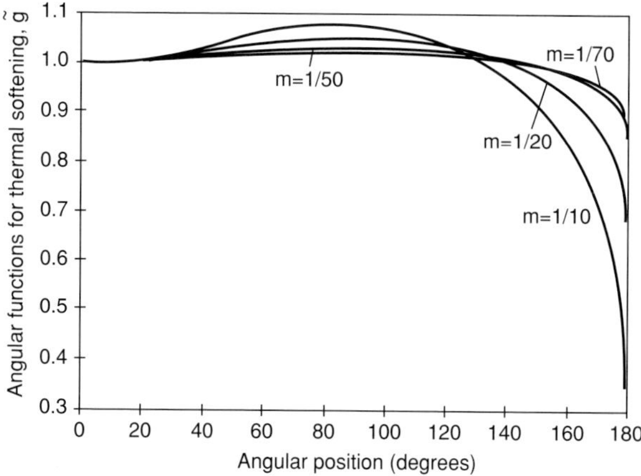

Figure 9.10. Thermal softening (or temperature since $a\theta = 1 - g$) relative to material ahead of the band. There is a ridge of hotter material at $\phi = 0$ and at $\phi = \pi$. (Reproduced from Chen and Batra 1999, with permission from Elsevier Science.)

and temperature. Also note that like the case of mode III propagation, when a material with a given strain rate sensitivity is calibrated to the conditions immediately ahead of the tip of the shear band, $G(0)$ is determined. There is no further arbitrariness in the solution because the speed of propagation, U, is determined by Equation (9.11).

As in the case of mode III motion the inertial solution, outlined above, must be joined to a core solution for a fully formed shear band near the line $\phi = \pi$. The details are entirely similar to the previous case and need not be repeated in full here.

9.3 The Leading Boundary Layer

Gioia and Ortiz (1996) have developed a two-dimensional theory for boundary layers and shear bands in thermoviscoplastic solids, which they have applied to an impact problem that simulates the Kalthoff type of experiment described in Chapter 8 and shown schematically in Figure 9.11. The material is assumed to be incompressible and rigid plastic with all constitutive functions being of a power law type, including the plastic flow potential, the heat capacity, and a generalized Fourier law for heat conduction. Plastic work is used as a fundamental variable in the flow law. In Cartesian tensor notation the equations for conservation of mass, momentum, and energy are as follows:

$$v_{i,i} = 0,$$

$$\rho(v_{i,t} + v_j v_{i,j}) = s_{ij,j} - p_{,i}, \tag{9.30}$$

$$\rho c(\theta_{,t} + v_i \theta_{,i}) = -q_{i,i} + \beta s_{ij} d_{ij},$$

where the comma denotes partial differentiation. The particle velocity is v, the deviatoric stress is s, the pressure is p, the temperature is θ, and the heat flux is q. The rate of deformation tensor is d with Cartesian components $d_{ij} = 1/2(v_{i,j} + v_{j,i})$. The deviatoric stress and heat flux are obtained from potentials

$$s_{ij} = \frac{\partial D}{\partial d_{ij}}, \quad q_i = -\frac{\partial E}{\partial g_i}, \tag{9.31}$$

where $g_i = \theta_{,i}$ is the temperature gradient. The constitutive laws for heat capacity, plastic flow, and heat conduction are assumed to be

$$c = c_0 \left(\frac{\theta}{\theta_0}\right)^q,$$

$$D = \frac{\sigma_0 \dot{\gamma}_0}{1 + m} \left(\frac{\dot{\gamma}}{\dot{\gamma}_0}\right)^{1+m} \left(\frac{w}{w_0}\right)^{\bar{n}} \left(\frac{\theta}{\theta_0}\right)^l, \tag{9.32}$$

$$E = \frac{q_0 g_0}{k + 1} \left(\frac{\theta}{\theta_0}\right)^p \left(\frac{g}{g_0}\right)^{k+1},$$

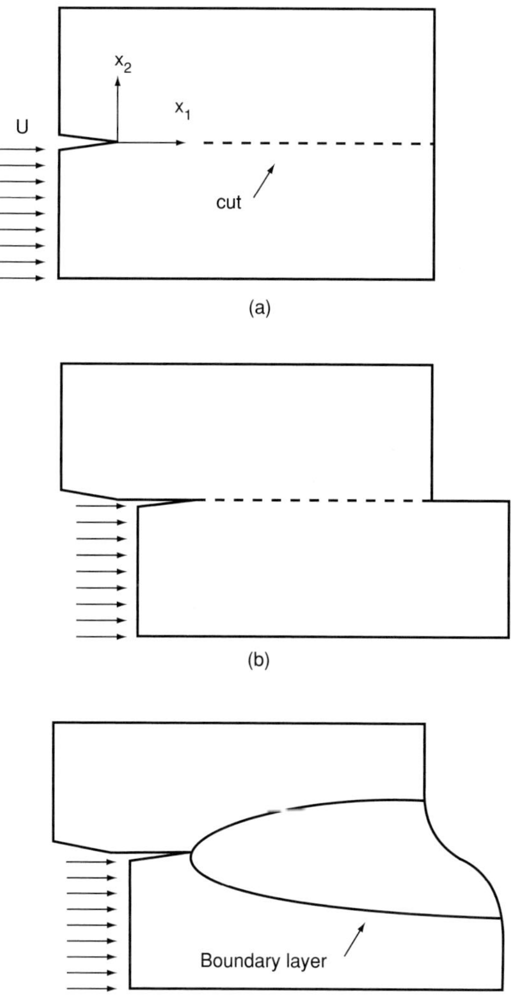

Figure 9.11. Thought experiment to illustrate the development of a boundary layer in a solid: panel a, impact on a prenotched plate; panel b, cut and translation of lower half develops incompatible flow; panel c, compatibility restored by development of a boundary layer between halves of the plate. (Reproduced from Gioia and Ortiz 1996, with permission from Elsevier Science.)

where $\bar{n} = n/(n+1)$; $k, l, m, n, p,$ and q are material constants; $c_0, \theta_0, \sigma_0, \dot{\gamma}_0,$ $w_0, q_0,$ and g_0 are constant reference values; and $\dot{\gamma} = (2d_{ij}d_{ij})^{1/2}, g = (g_i g_i)^{1/2},$ and the plastic work w is determined from the evolution equation

$$\dot{w} \equiv w_{,t} + v_i w_{,i} = s_{ij} d_{ij}. \tag{9.33}$$

For two-dimensional applications the range of the indices i and j is restricted to 1, 2 only.

The power law structure assumed above for the constitutive laws is convenient both for fitting material data and for a scaling analysis based on the notion that solutions have the structure of a boundary layer. The basic assumption is that if the solution varies appreciably in the x_1 direction over a distance L, then in the x_2 direction the solution varies appreciably over the distance εL, where ε is a small parameter, $\varepsilon \ll 1$. To maintain incompressibility the velocity field is generated by a stream function, $v_1 = \psi_{,2}$, $v_2 = -\psi_{,1}$. Scaled variables are introduced next,

$$x_1' = x_1/L, \quad x_2' = x_2/\varepsilon L, \quad t' = Ut/L, \quad \psi' = \psi/\varepsilon U L, \tag{9.34}$$

where U is a velocity that is characteristic of the free flow. The velocity field now takes the form

$$v_1 = U v_1' = U \psi_{,2'}', \quad v_2 = \varepsilon U v_2' = -\varepsilon U \psi_{,1'}'. \tag{9.35}$$

In this equation v_1' and v_2' are $\mathcal{O}(1)$ quantities and the notation $(\cdot)_{,i'}$ denotes partial differentiation with respect to x_i'. To leading order in ε only one component of the stretching tensor need be retained:

$$d_{12} \sim \frac{1}{2\varepsilon} \frac{U}{L} v_{1,2'}'. \tag{9.36}$$

This result is reminiscent of the scaling used by DiLellio and Olmstead, as described in Chapter 7, but as we shall soon see, in fact, it is quite different.

The scaling adopted so far also ensures that the dominant component of stress is s_{12}, that the plastic stress power is dominated by the term $2s_{12}d_{12}$, and that $q_{2,2}$ is the dominant term in the divergence of heat flux.

Temperature and plastic work are also scaled by undetermined powers of ε. These unknown exponents are to be determined by the standard method of balancing powers in the governing equations so as to retain only the most dominant terms. When applied to the energy equation, however, this procedure leads to a choice between two alternatives, either of which is possible. If conduction is dominant, the exponents are such that the principal terms for plastic stress power and heat flux balance as ε vanishes. If convection is dominant, the principal terms for plastic stress power and material rate of temperature increase must balance as ε vanishes. These two choices are closely related to the steady and adiabatic cases that have been encountered so many times previously.

When these scaling procedures are applied to the momentum equation, the result is

$$v_{1,t'}' + v_1' v_{1,1'}' + v_2' v_{1,2'}' = \varepsilon^{\alpha-1} \frac{S}{\rho U^2} s_{12,2'}' - p_{,1'}', \tag{9.37}$$

where $\alpha < 1$ has been determined from the constitutive exponents k, l, m, n, p, and q as required by the scaling procedures, and S is a characteristic stress. The authors then argue that because $\alpha = 1$ is incompatible with previous scaling results, and because all terms in the equation should be $\mathcal{O}(1)$, it follows that $\varepsilon^{\alpha-1}(S/\rho U^2)$ must itself be $\mathcal{O}(1)$. The small parameter ε has not yet been defined, so at this point it is chosen in such a way that $\varepsilon^{\alpha-1}(S/\rho U^2) = 1$ or

$$\varepsilon = \mathcal{R}^{-1/(1-\alpha)}, \quad \mathcal{R} = \rho U^2/S. \tag{9.38}$$

The number $\mathcal{R}$ is a generalized Reynolds number, which must be large in order to make the parameter ε small. This situation is very different from any encountered previously in which the small expansion parameter has usually been the strain rate sensitivity, m. For the example presented below, the impact velocity chosen to simulate a Kalthoff type of experiment is 544 m/s with a Reynolds number of 10. This value for the impact velocity is much higher than many experimental values of interest, as discussed in the last chapter. In fact, many situations there have impact velocities that are more than an order of magnitude smaller than 544 m/s and therefore correspond to Reynolds numbers that are two or three orders of magnitude smaller, that is, $\mathcal{R} < \mathcal{O}(10^{-1})$. Therefore, these cases actually lie in a regime of small Reynolds numbers. Nevertheless, the solutions found by Gioia and Ortiz with their scheme for scaling the equations are very interesting and shed much light on the way that an adiabatic shear band may fit in with a global solution to a boundary value problem.

As noted above, the scaling reduces the governing equations to one of two forms: either conduction or convection may dominate the energy equation in the limit that $\varepsilon \to 0$ or equivalently $\mathcal{R} \to \infty$, and for each case α may be calculated to be a definite number that may be determined from the constitutive exponents. However, for the actual metals considered, namely Al, Fe, and Cu, only the choice of convection is fully consistent with the ordering of terms in the energy equation. Nevertheless, because of stability considerations it always turns out that a thin conduction layer is required at the very core of the convective boundary layer. This situation is essentially the same as that described in the first two sections of this chapter. The outer solution is dominated by inertia and convection, but for real materials this solution is singular at its core; the inner layer at the core of the shear band, which may be connected to the outer solution in a systematic way, is dominated by thermal conduction and is always stable, that is, nonsingular.

The boundary layer equations, which by construction have three independent variables – two in space and one in time – may be reduced to one or two independent variables by appropriately chosen similarity transformations. For full details the reader is referred to the original publication (Gioia and Ortiz 1996) but in outline the procedure is as follows. First, with a slight variation in the choice of scaled variables,

$$\tilde{x}_1 = x_1/L, \quad \tilde{x}_2 = x_2/L, \quad \tilde{t} = Ut/L, \quad \tilde{\psi} = \psi/UL,$$
$$\tilde{v}_1 = v_1/U, \quad \tilde{v}_2 = v_2/U, \quad \tilde{\theta} = \theta/T, \quad \tilde{w} = w/W, \quad \tilde{p} = p/\rho U^2, \tag{9.39}$$

the boundary layer equations for conservation of mass and momentum become

$$\frac{\partial \tilde{v}_1}{\partial \tilde{x}_1} + \frac{\partial \tilde{v}_2}{\partial \tilde{x}_2} = 0,$$

$$\frac{\partial \tilde{v}_1}{\partial \tilde{t}} + \tilde{v}_1 \frac{\partial \tilde{v}_1}{\partial \tilde{x}_1} + \tilde{v}_2 \frac{\partial \tilde{v}_1}{\partial \tilde{x}_2} = \frac{1}{\mathcal{R}} \frac{\partial \tilde{s}_{12}}{\partial \tilde{x}_2} + \frac{\partial \tilde{p}}{\partial \tilde{x}_1}, \tag{9.40}$$

$$\frac{\partial \tilde{p}}{\partial \tilde{x}_2} = 0,$$

the constitutive law for stress becomes

$$\tilde{s}_{12} = \left| \frac{\partial \tilde{v}_1}{\partial \tilde{x}_2} \right|^{-1+m} \frac{\partial \tilde{v}_1}{\partial \tilde{x}_2} \tilde{w}^{\bar{n}} \tilde{\theta}^l, \tag{9.41}$$

the transport equation for plastic work becomes

$$\frac{\partial \tilde{w}}{\partial \tilde{t}} + \tilde{v}_1 \frac{\partial \tilde{w}}{\partial \tilde{x}_1} + \tilde{v}_2 \frac{\partial \tilde{w}}{\partial \tilde{x}_2} = \mathcal{P} \tilde{s}_{12} \frac{\partial \tilde{v}_1}{\partial \tilde{x}_2}, \quad \mathcal{P} = \frac{S}{W}, \tag{9.42}$$

and the equation for conservation of energy when conduction dominates becomes

$$-\frac{\partial \tilde{q}_2}{\partial \tilde{x}_2} = \mathcal{T}_d \tilde{s}_{12} \frac{\partial \tilde{v}_1}{\partial \tilde{x}_2},$$

$$\tilde{q}_2 = \tilde{\theta}^p \left| \frac{\partial \tilde{\theta}}{\partial \tilde{x}_2} \right|^{k-1} \frac{\partial \tilde{\theta}}{\partial \tilde{x}_2}, \quad \mathcal{T}_d = \frac{\beta S(U/L)}{(B/L)T^p(T/L)^k}. \tag{9.43}$$

If convection dominates, the equation for conservation of energy becomes

$$\tilde{\theta}^q \left(\frac{\partial \tilde{\theta}}{\partial \tilde{t}} + \tilde{v}_1 \frac{\partial \tilde{\theta}}{\partial \tilde{x}_1} + \tilde{v}_2 \frac{\partial \tilde{\theta}}{\partial \tilde{x}_2} \right) = \mathcal{T}_v \tilde{s}_{12} \frac{\partial \tilde{v}_1}{\partial \tilde{x}_2}, \quad \mathcal{T}_v = \frac{\beta S(U/L)}{\rho C T^{q+1}(U/L)}. \tag{9.44}$$

These equations may be reduced to coupled ordinary differential equations if the flow is assumed to be steady, the similarity variable

$$\varsigma = \tilde{x}_2 \tilde{x}_1^{-a} \tag{9.45}$$

is introduced, and the fields are expressed as

$$\tilde{\psi} = \tilde{x}_1^a f(\varsigma), \quad \tilde{\theta} = \tilde{x}_1^b g(\varsigma), \quad \tilde{w} = \tilde{x}_1^c h(\varsigma). \tag{9.46}$$

The velocity fields and the shear stress become

$$\tilde{v}_1 = f', \quad \tilde{v}_2 = -a\tilde{x}_1^{a-1}(f - f'\varsigma), \quad \tilde{s}_{12} = \tilde{x}_1^d \tau \operatorname{sgn}(f''),$$
$$\tau = |f''|^m g^l h^{\bar{n}}, \quad d = -am + cn + bl. \tag{9.47}$$

After all these transformations for the case that is dominated by convection, the coupled system reduces to

$$-aff'' = \frac{1}{\mathcal{R}}\tau\left(m\frac{f'''}{|f''|} + \bar{n}\frac{h'}{h} + l\frac{g'}{g}\right),$$
$$chf' - afh' = \mathcal{P}\tau|f''|, \tag{9.48}$$
$$bg^{1+q}f' - afg^q g' = \mathcal{T}_v\tau|f''|,$$

with characteristic exponents

$$a = 1/(1 + m), \quad b = c = 0. \tag{9.49}$$

This case is closely related to those discussed in Sections 9.1 and 9.2 of this chapter, as may be seen by examining the expression for the dominant component of strain rate

$$\tilde{v}_{1,2} = x_1^{-a} f'' = (r\cos\theta)^{-1/(1+m)} f''. \tag{9.50}$$

Note that the singular behavior in the radius r is the same here as in those previous cases. However, in this case, the radius is measured from a fixed origin, whereas in the previous cases it was measured from a moving point.

For the case that is dominated by conduction, the first two equations of (9.48) remain the same, but the energy equation and the characteristic exponents take different forms. These may be found in the paper.

Figure 9.12 shows an example worked out by Gioia and Ortiz. The velocity and stress are continuous at the origin, but the temperature and plastic work become singular. The singularity is merely an artifact of the convective solution, which must be eliminated by a nested conduction layer, as indicated in the temperature profile. In this example the striking velocity is 544 m/s with a Reynolds number of 10, as stated previously. The horizontal velocity transitions from an asymptotic value equal to the impact velocity in the bottom half of the figure to zero in the top half. The shear stress is asymmetric because of the need to match outer boundary conditions and let the solution progress inward from both sides to the horizontal axis. However, it is most intense at the origin, which is the center of the boundary layer, and the temperature and plastic work are sharply peaked at the origin. In addition, both the stress and the conduction temperature fields vary with the coordinate x_1, as well as ς, with stress showing a decrease downstream and temperature showing an increase.

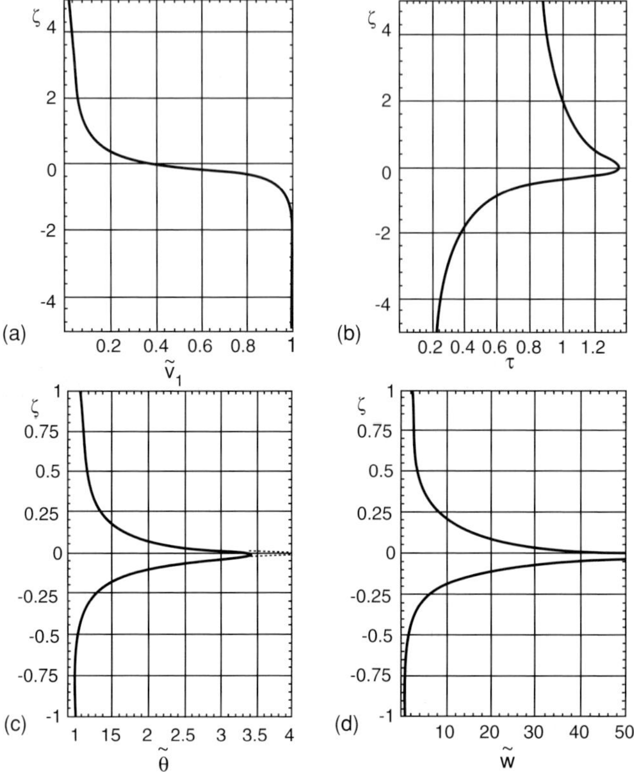

Figure 9.12. Steady boundary layer in a Cu plate showing the distribution of velocity (panel a), shear stress (panel b), temperature (panel c), and plastic work across the layer (panel d). (Reproduced from Gioia and Ortiz 1996, with permission from Elsevier Science.)

The full convective boundary layer equations, (9.40)–(9.42) and (9.44), also have transient similarity solutions. In this case two variables are introduced,

$$\varsigma = \tilde{x}_2/\tilde{x}_1^a, \quad \xi = \tilde{t}/\tilde{x}_1, \tag{9.51}$$

together with the representation

$$\tilde{\psi} = \tilde{x}_1^a f(\varsigma, \xi), \quad \tilde{\theta} = \tilde{x}_1^b g(\varsigma, \xi), \quad \tilde{w} = \tilde{x}_1^c h(\varsigma, \xi). \tag{9.52}$$

Choosing exponents so as to balance out powers of $\tilde{x}_1$ gives the same values as in the steady case, Equation (9.49). After use of the integrating factor $(1 - \xi f_{,\varsigma})$, the convective equations may now be recast as

$$\left(\frac{f_{,\xi}}{1 - \xi f_{,\varsigma}}\right)_{,\varsigma} = \frac{\mathcal{R}^{-1} s_{,\varsigma} + a f f_{,\varsigma\varsigma}}{(1 - \xi f_{,\varsigma})^2},$$

$$g_{,\xi} = \frac{g^{-q} \mathcal{T}_v s f_{,\varsigma\varsigma} + (af - \xi f_{,\xi}) g_{,\varsigma}}{1 - \xi f_{,\varsigma}}, \tag{9.53}$$

$$h_{,\xi} = \frac{\mathcal{P} s f_{,\varsigma\varsigma} + (af - \xi f_{,\xi}) h_{,\varsigma}}{1 - \xi f_{,\varsigma}},$$

$$s = |f_{,\varsigma\varsigma}|^m g^l h^{\bar{n}} \mathrm{sgn}(f_{,\varsigma\varsigma}).$$

These equations show a possible transition to the steady equations if the material is capable of sustaining a steady boundary layer. This may be seen from the following considerations. Because $f_{,\varsigma} = \tilde{v}_1$ and $0 < \tilde{v}_1 < 1$, for every ς there is a $\xi^*(\varsigma)$ such that

$$1 - \xi^* f_{,\varsigma}(\varsigma, \xi^*) = 0. \tag{9.54}$$

At this ς-dependent value of ξ^*, the denominators on both sides of the first equation of (9.53) vanish. If the equation is to be well behaved at this point, the numerators on both sides must also vanish. In turn this requires that $f_{,\xi} = 0$, which indicates steady behavior for f. The numerator on the right side then reduces to the first equation of (9.48), which is the steady equation for balance of momentum. The numerators of the second and third equations of (9.53) also must vanish. However, these are just the other steady equations from (9.48), which suggests that g and h are also steady for greater values of ξ. Furthermore, because of the definition of the similarity variable $\xi = \tilde{t}/\tilde{x}_1$, there is a definite time at which the steady state may be achieved at a given point:

$$\tilde{t}^* = \tilde{x}_1 \xi^* \left(\tilde{x}_2 \tilde{x}_1^{-a}\right). \tag{9.55}$$

Figure 9.13 shows an example of the evolution of the boundary between transient and steady state for the same example as shown in Figure 9.12. Notice, however, that at large, negative values of $\tilde{x}_2$, the arrival time of the steady state is $\tilde{t} = \tilde{x}_1$, which is just the arrival time of the impactor itself. It appears that points at depths greater than the depth of indentation can never achieve a steady state. In fact, the moving boundary introduced by the impactor has not been explicitly recognized in posing the problem so the interpretation of this point must remain unclear. If the problem were posed in material rather than spatial coordinates, such ambiguities should disappear.

To investigate the speed at which a shear band propagates, the authors propose using a somewhat arbitrary criterion. At some critical level of plastic work, $\tilde{w}_c$, it is assumed that the work hardening exponent suddenly reduces so that

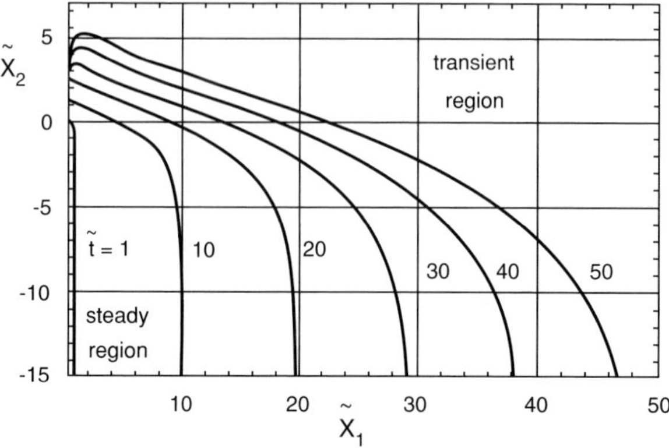

Figure 9.13. Evolution of the steady-state boundary in a plate that has been struck on the lower half. (Reproduced from Gioia and Ortiz 1996, with permission from Elsevier Science.)

the material passes from a stable to an unstable condition. If this is the case, then for each instant of time there is a contour at which this value is reached:

$$\tilde{w}\left(\tilde{x}_2/\tilde{x}_1^a, \tilde{t}/\tilde{x}_1\right) = \tilde{w}_c. \tag{9.56}$$

In particular, the tip of the contour, where $\tilde{x}_2 = 0$, satisfies

$$\tilde{w}(0, \tilde{t}/\tilde{x}_1) = \tilde{w}_c, \tag{9.57}$$

which implies that there is a critical value of $\xi = \tilde{t}/\tilde{x}_1$ that depends on $\tilde{w}_c$. In turn this implies that for a given critical value of plastic work, there is a constant speed of propagation for the tip of the shear band:

$$\tilde{V} = 1/[\xi_c(\tilde{w}_c)]. \tag{9.58}$$

Figure 9.14 shows contours of constant plastic work at several instants of time, and clearly the tip speed for each value is constant. Note that the transverse axis in the figure is plotted at twice the scale of the horizontal axis so the actual elongation is even more exaggerated than appears in the figure. The contours of constant plastic work and also of constant temperature (not shown) are long narrow plumes that shoot downstream ahead of the impactor. The rate of deformation, however, although also propagating downstream in a long plume, tends to reach a maximum distance and then to retreat. This occurs because it has the representation $\tilde{d}_{12} = \tilde{x}_1^{-a} f_{,\varsigma\varsigma}(\varsigma, \xi)$. Thus, at the tip of the curve where $\tilde{d}_{12} = \tilde{d}$ and $\varsigma = 0$, the relation between distance and time is

$$\bar{d} = \tilde{x}_1^{-a} f_{,\varsigma\varsigma}(0, \tilde{t}/\tilde{x}_1). \tag{9.59}$$

Therefore, $\tilde{x}_1/\tilde{t} = \text{const}$ is not a solution.

The complete paper is far richer in detail than can be presented here.

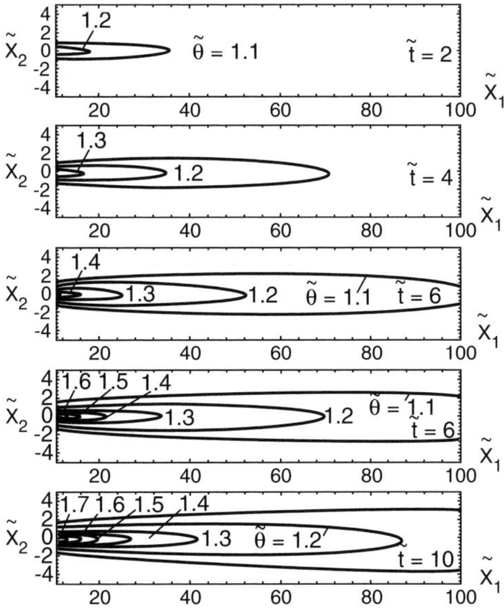

Figure 9.14. Evolution of the temperature field in the boundary layer that precedes the fully developed shear band. (Reproduced from Gioia and Ortiz 1996, with permission from Elsevier Science.)

9.4 Summary

Three two-dimensional analyses of adiabatic shear bands have been summarized in this chapter. Local analyses for mode II and mode III shearing were constructed for the tip of a steadily translating shear band by assuming a power law variation in the radius, measured from the tip. In both cases this procedure leads to a set of nonlinear ordinary differential equations with the polar angle around the tip as an independent variable. Strain rate sensitivity is the only physical constant that remains in the ODEs in both cases, because all other constants scale out as nondimensional multiplying factors for the dependent variables. These factors are essentially the same for similar physical quantities in both mode II and mode III with the quantity $a\kappa_0/\rho c$ playing a prominent role, as it has throughout this book. This is the nondimensional coefficient of thermal softening, which more generally may be written $-s_\theta/\rho c$, that is, the partial derivative of the flow stress with respect to temperature, divided by density and heat capacity. Furthermore, the power of the radial variation is also the same for similar physical quantities in both mode II and mode III, as is the theoretical unconstrained speed of propagation.

Another analysis, using only power law constitutive relations, adapts classical boundary layer techniques from fluid mechanics to examine certain aspects of the flow fields that develop in the Kalthoff experiment. The analysis is very rich in detail that is not easily summarized, but one striking result clearly shows the long plume of plastic work and higher temperature that rapidly propagates out ahead of the developing shear band.

So far, two-dimensional analyses have been somewhat limited in scope and have concentrated only on particular aspects of shear band formation and propagation. Much still remains to be learned, especially about band propagation and its connection to the larger scale deformation within which it is embedded.

Bibliography

Adams, B. L., Boehler, J. P., Guidi, M., and Onat, E. T. (1992). Group theory and representation of microstructure and mechanical behavior of polycrystals. *Journal of the Mechanics and Physics of Solids*, **40**, 723–737.

Anand, L. (1985). Constitutive equations for hot-working of metals. *International Journal of Plasticity*, **1**, 213–231.

Anand, L. and Brown, S. (1987). Constitutive equations for large deformations of metals at high temperature. In *Constitutive Models of Deformation*, ed. J. Chandra and R. Srivastav, pp. 1–26. Philadelphia: Society of Industrial and Applied Mathematics.

Anand, L., Kim, K. H., and Shawki, T. G. (1987). Onset of shear localization in viscoplastic solids. *Journal of the Mechanics and Physics of Solids*, **35**, 407–429.

Aravas, N., Kim, K.-S., and Leckie, F. A. (1990). On the calculation of the stored energy of cold work. *Journal of Engineering Materials and Technology*, **112**, 465–470.

Armstrong, R. W., Batra, R. C., Meyers, M. A., and Wright, T. W. (eds.) (1994). Shear Instabilities and Viscoplasticity Theories. Mechanics of Materials, **17**, 83–328.

Asaro, R. J. (1982). Micromechanics of crystals and polycrystals. In *Advances in Applied Mechanics*, Vol. 23, eds. T. Y. Wu and J. W. Hutchinson, pp. 1–113. New York: Academic Press.

Ashby, M. F. (1992). *Materials Selection in Mechanical Design*. Oxford: Pergamon Press.

Bai, Y. (1982). Thermo-plastic instability in simple shear. *Journal of the Mechanics and Physics of Solids*, **30**, 195–207.

(1989). Evolution of thermo-visco-plastic shearing. In *Mechanical Properties of Materials at High Rates of Strain 1989*, ed. J. Harding, pp. 99–110, Institute of Physics Conference Series No. 102. Bristol: Institute of Physics.

Bai, Y. and Dodd, B. (1992). *Adiabatic Shear Localization*. Oxford: Pergamon Press.

Bammann, D. J. (1990). Modeling the temperature and strain rate dependent large deformation of metals. *Applied Mechanics Reviews*, **43**, S312–S319.

Bammann, D. J. and Johnson, G. C. (1987). On the kinematics of finite-deformation plasticity. *Acta Mechanica*, **70**, 1–13.

Batra, R. C. (1988). Steady state penetration of thermoviscoplastic targets. *Computational Mechanics*, **3**, 1–12.

(1998). Numerical solutions of initial-boundary-value problems. In *Localization and Fracture Phenomena in Inelastic Solids*, ed. P. Perzyna, pp. 301–389. New York: Springer.

Batra, R. A., Rajapakse, Y. D. S., and Zbib, H. M. (eds.) (1994). *Material Instabilities: Theory and Applications*. New York: ASME Press.

Batra, R. C. and Gummalla, R. R. (2000). Effect of material and geometric parameters on deformations near the notch-tip of a dynamically loaded prenotched plate. *International Journal of Fracture*, **101**, 99–140.

Batra, R. C. and Ravinsankar, M. V. S. (2000). Three-dimensional simulation of the Kalthoff experiment. *International Journal of Fracture*, **105**, 161–186.

Batra, R. C. and Wright, T. W. (1986). Steady state penetration of rigid perfectly plastic targets. *International Journal of Engineering Science*, **24**, 41–54.

Beatty, J. H., Meyer, L. W., Meyers, M. A., and Nemat-Nasser, S. (1992). Formation of controlled adiabatic shear bands in AISI 4340 high strength steel. In *Shock Wave and High-Strain-Rate Phenomena in Materials*, ed. M. A. Meyers, L. E. Murr, and K. P. Staudhammer, pp. 645–656. New York: Marcel Dekker.

Bell, J. F. (1974). The experimental foundations of solid mechanics. In *Mechanics of Solids*, Vol. 1, ed. C. Truesdell, p. 464. Berlin: Springer-Verlag.

Bender, C. M. and Orszag, S. A. (1978). *Advanced Mathematical Methods for Scientists and Engineers*. New York: McGraw-Hill.

Bever, M. B., Holt, D. L., and Titchener, A. L. (1973). The stored energy of cold work. *Progress in Material Science*, **17**, 1–190.

Bodner, S. R. (1987). Review of a unified elastic-viscoplastic theory. In *Unified Constitutive Theories for Creep and Plasticity*, ed. A. K. Miller, pp. 273–301. New York: Elsevier.

Bodner, S. R. and Lindenfeld, A. (1995). Constitutive modeling of the stored energy of cold work under cyclic loading. *European Journal of Mechanics*, **A14**, 333–348.

Brown, S. B., Kim, K. H., and Anand, L. (1989). An internal variable constitutive model for hot-working of metals. *International Journal of Plasticity*, **5**, 95–130.

Campbell, J. D. and Dowling, A. R. (1970). The behavior of materials subjected to dynamic incremental shear loading. *Journal of the Mechanics and Physics of Solids*, **18**, 43–63.

Carlson, D. E. (1972). Linear thermoelasticity. In *Encyclopedia of Physics*, Vol. VIa/2, ed. C. Truesdell. Berlin: Springer-Verlag.

Chadwick, P. (1976). *Continuum Mechanics*. New York: Halstead Press, Wiley.

Chen, HTz (1988). Stability conditions for steady shearing and stability conditions for simple shear under mixed thermal boundary conditions. Baltimore, MD: PhD dissertation, The Johns Hopkins University.

Chen, HTz, Douglas, A. S., and Malek-Madani, R. (1989). An asymptotic stability condition for inhomogeneous simple shear. *Quarterly of Applied Mathematics*, **47**, 247–262.

Chen, L. and Batra, R. C. (1999). The asymptotic structure of a shear band in mode-II deformations. *International Journal of Engineering Science*, **37**, 895–919.

Chen, Y.-J., Meyers, M. A., and Nesterenko, V. F. (1999). Spontaneous and forced localization in high-strain-rate deformation of tantalum. *Materials Science and Engineering*, **A268**, 70–82.

Chichili, D. R. (1997). High-strain-rate deformation mechanisms and adiabatic shear localization in alpha-titanium. Baltimore, MD: PhD dissertation. The Johns Hopkins University.

Chichili, D. R. and Ramesh, K. T. (1999). Recovery experiments for adiabatic shear localization: A novel experimental technique. *Journal of Applied Mechanics*, **66**, 10–20.

Chichilli, D. R., Ramesh, K. T., and Hemker, K. J. (1998). The high-strain-rate response of alpha-titanium: Experiments, deformation mechanisms, and modeling. *Acta Materialia*, **46**, 1025–1043.

Clarebrough, L. M., Hargreaves, M. E., and Loretto, M. H. (1962). Electrical resistivity of dislocations in face-centered cubic metals. *Philisophical Magazine*, **7**, 115–120.

Clarebrough, L. M., Hargreaves, M. E., Michel, D., and West, G. W. (1952). The determination of the energy stored in a metal during plastic deformation. *Proceedings of the Royal Society of London*, **A215**, 507–524.

Clarebrough, L. M., Hargreaves, M. E., and West, G. W. (1955). The release of energy during annealing of deformed metals. *Proceedings of the Royal Society of London*, **A232**, 252–270.

   (1956). Density changes during the annealing of deformed nickel. *Philisophical Magazine*, **1**, 528–536.

   (1957). The density of dislocations in compressed copper. *Acta Metallurgica*, **5**, 738–740.

Cleja-Ţigoiu, S. and Soós, E. (1990). Elastoviscoplastic models with relaxed configurations and internal state variables. *Applied Mechanics Reviews*, **43**, 131–151.

Clifton, R. J. (1980). Adiabatic shear banding. Chapter 8 in *Materials Response to Ultra-High Loading Rates*, NMAB-356. Washington, DC: National Academy of Sciences.

Clifton, R. J. and Klopp, R. W. (1985). Pressure-shear plate impact testing. In *ASM Metals Handbook*, 9th ed., Vol. 8, *Mechanical Testing*, pp. 230–238. Materials Park, OH: ASM International.

Coates, R. S. and Ramesh, K. T. (1991). The rate-dependent deformation of a tungsten heavy alloy. *Materials Science and Engineering*, **A145**, 159–166.

Costin, L. S., Crisman, E. E., Hawley, R. H., and Duffy, J. (1979). On the localization of plastic flow in mild steel tubes under dynamic torsional loading. In *Mechanical Properties at High Rates of Strain 1979*, Institute of Physics Conference Series No. 47, ed. J. Harding, pp. 90–110. Bristol: Institute of Physics.

Cowie, J. G., Azrin, M., and Olson, G. B. (1989). Microvoid formation during shear deformation of ultrahigh strength steels. *Metallurgical Transactions*, **A20**, 143–153.

Curran, D. R., Seaman, L., and Shockey, D. A. (1987). Dynamic failure of solids. *Physics Reports*, **147**, 253–388.

DiLellio, J. A. and Olmstead, W. E. (1997a). Temporal evolution of shear band thickness. *Journal of the Mechanics and Physics of Solids*, **45**, 345–359.

   (1997b). Shear band formation due to a thermal flux inhomogeneity. SIAM *Journal on Applied Mathematics*, **57**, 959–971.

   (1998). Numerical solutions of shear localization in a finite slab. *Mechanics of Materials*, **29**, 71–80.

Dodd, B. and Bai, Y. (1985). Width of adiabatic shear bands. *Materials Science and Technology*, **1**, 38–40.

Dormeval, R. (1988). Adiabatic shear phenomena. In *Impact Loading and Dynamic Behavior of Materials*, Vol. 1, eds. C. Y. Chiem, H.-D. Kunze, and L. W. Meyer, pp. 43–56. Germany: DGM Informationsgesellschaft.

Duffy, J. (1980). The J. D. Campbell memorial lecture: testing techniques and material behavior at high rates of strain. In *Mechanical Properties at High Rates of Strain 1979*, Institute of Physics Conference Series No. 47, ed. J. Harding, pp. 1–15. Bristol: Institute of Physics.

Duffy, J. and Chi, Y. C. (1992). On the measurement of local strain and temperature during the formation of adiabatic shear bands. *Materials Science and Engineering*, **A157**, 195–210.

Erlich, D. C., Seaman, L., and Shockey, D. A. (1980). Development of a computational shear band model. Final Report for Contract DAADOS-76-C-0762. Aberdeen Proving Ground, MD: U.S. Army Ballistic Research Laboratory.

Farren, W. S. and Taylor, G. I. (1925). The heat developed during plastic extension of metals. *Proceedings of the Royal Society of London*, **A107**, 422–451.

Farshisheh, F. and Onat, E. T. (1974). Representation of elastoplastic behavior by means of state variables. In *Proceedings of the International Symposium on Plasticity*, ed. A. Sawczuk, pp. 89–115. Leyden: Noordhoff.

Field, J. E., Walley, S. M., Bourne, N. K., and Huntley, J. M. (1994). Experimental methods at high rates of strain. *Journal de Physique IV*, Colloque C8, supplement au Journal de Physique III, C8-3–C8-22.

Flemming, R. P., Olmstead, W.-E., and Davis, S. H. (2000). Shear localization with an Arrhenius flow law. *SIAM Journal on Applied Mathematics*, **60**, 1867–1886.

Follansbee, P. S. (1985). The Hopkinson bar. In *ASM Metals Handbook*, 9th ed., Vol. 8, *Mechanical Testing*, pp. 198–203. Materials Park, OH: ASM International.

   (1989). Analysis of the strain-rate sensitivity at high strain rates in fcc and bcc metals. In *Mechanical Properties of Materials at High Rates of Strain 1989*, Institute of Physics Conference Series No. 102, ed. J. Harding, pp. 213–220. Bristol: Institute of Physics.

Follansbee, P. S. and Frantz, C. (1983). Wave propagation in the split-Hopkinson pressure bar. *Journal of Engineering Materials and Technology*, **105**, 61–66.

Follansbee, P. S. and Kocks, U. F. (1988). A constitutive description of copper based on the use of the mechanical threshold stress as an internal state variable. *Acta Metallurgica*, **36**, 81–93.

Follansbee, P. S., Regazzoni, G., and Kocks, U. F. (1984). The transition to drag-controlled deformation in copper at high strain rates. In *Mechanical Properties at High Rates of Strain 1984*, Institute of Physics Conference Series No. 70, ed. J. Harding, pp. 71–80. Bristol: Institute of Physics.

Gilat, A. and Wu, X. (1994). Elevated-temperature testing with the torsional split Hopkinson bar. *Experimental Mechanics*, **34**, 166–170.

Gioia, G. and Ortiz, M. (1996). The two-dimensional structure of dynamic boundary layers and shear bands in thermoviscoplastic solids. *Journal of the Mechanics and Physics of Solids*, **44**, 251–292.

Giovanola, J. H. (1988). Adiabatic shear banding under pure shear loading. Part 1: Direct observation of strain localization and energy-dissipation measurements. *Mechanics of Materials*, **7**, 59–71.

Glimm, J., Plohr, B. J., and Sharp, D. H. (1993). A conservative formulation for large deformation plasticity. *Applied Mechanics Reviews*, **46**, 519–526.

   (1995). Tracking of shear bands. I. The one-dimensional case. SUNYSB-AMS-95-04. Stony Brook, NY: State University of New York.

Gorham, D. A., Pope, P. H., and Field, J. E. (1992). An improved method for compressive stress-strain measurements at very high strain rates. *Proceedings of the Royal Society of London*, **A438**, 153–170.

Grady, D. E., Asay, J. R., Rohde, R. W., and Wise, J. L. (1983). Microstructure and mechanical properties of precipitation hardened aluminum under high rate deformation. In *Material Behavior Under High Stress and Ultrahigh Loading Rates*, 29th Sagamore Conference, eds. J. Mescal and V. Weiss, pp. 161–196. New York: Plenum Press.

Grady, D. E. and Kipp, M. E. (1987). The growth of unstable thermoplastic shear with application to steady-wave shock compression in solids. *Journal of the Mechanics and Physics of Solids*, **35**, 95–120.

Grebe, H. A., Pak, H.-R., and Meyers, M. A. (1985). Adiabatic shear localization in titanium and Ti-6 pct Al-4 pct V alloy. *Metallurgical Transactions*, **16A**, 761–775.

Hale, J. K. (1980). *Ordinary Differential Equations*, 2nd ed. Malabar, FL: Krieger.

Hartley, K. A., Duffy, J., and Hawley, R. H. (1985). The torsional Kolsky (split Hopkinson) bar. In *ASM Metals Handbook*, 9th ed., Vol. 8, *Mechanical Testing*, pp. 218–228. Materials Park, OH: ASM International.

Hartley, K. A., Duffy, J., and Hawley, R. H. (1987). Measurement of the temperature profile during shear band formation in steels deforming at high-strain rates. *Journal of the Mechanics and Physics of Solids*, **35**, 283–301.

Hartmann, K.-H., Kunze, H.-D., and Meyer, L. W. (1981). Metallurgical effects on impact loaded materials. In *Shock Waves and High-Strain-Rate Phenomena in Metals*, eds. M. A. Meyers and L. E. Murr, pp. 325–337. New York: Plenum Press.

Havner, K. S. (1992). *Finite Plastic Deformation of Crystalline Solids*. Cambridge: Cambridge University Press.

Hill, R. and Rice, J. R. (1972). Constitutive analysis of elastic-plastic crystals at arbitrary strain. *Journal of the Mechanics and Physics of Solids*, **20**, 401–413.

Hodowany, J., Ravichandran, G., Rosakis, A. J., and Rosakis, P. (2000). Partition of plastic work into heat and stored energy in metals. *Experimental Mechanics*, **40**, 113–123.

Holden, A. N. (1958). *Physical Metallurgy of Uranium*. Reading, MA: Addison-Wesley.

Jeffreys, H. (1962). *Asymptotic Approximations*. London: Oxford University Press.

Johnson, G. R. and Cook, W. H. (1983). A constitutive model and data for metals subjected to large strains, high strain rates, and high temperatures. In *Proceedings of the 7th International Symposium on Ballistics*, pp. 541–547. Organized under the auspices of the Royal Institution of Engineers (Klvl), Division for Military Engineering in cooperation with the American Defense Preparedness Association. The Hague.

Johnson, G. R., Hoegfeldt, J. M., Lindholm, U. S., and Nagy, A. (1983). Response of various metals to large torsional strains over a large range of strain rates – Part 1: Ductile metals. *Journal of Engineering Materials and Technology*, **105**, 42–47.

Johnson, W. (1987). Henry Tresca as the originator of adiabatic heat lines. *International Journal of Mechanical Sciences*, **29**, 301–310.

Kalthoff, J. F. (1987). Shadow optical analysis of dynamic shear fracture. SPIE, Vol. 814. *Photo Mechanics and Speckle Metrology*, 531–538.

  (1990). Transition in the failure behavior of dynamically shear loaded cracks. *Applied Mechanics Reviews*, **43**, S247–S250.

  (2000). Modes of dynamic shear failure in solids. *International Journal of Fracture*, **101**, 1–31.

Kalthoff, J. F. and Winkler, S. (1988). Failure mode transitions at high rates of shear loading. In *Impact Loading and Dynamic Behavior of Materials*, Vol. 1, eds. C. Y. Chiem, H.-D. Kunze, and L. W. Meyer, pp. 185–195. Germany: DGM Informationsgesellschaft mbH.

Kamlah, M. and Haupt, P. (1997). On the macroscopic description of stored energy and self heating during plastic deformation. *International Journal of Plasticity*, **13**, 893–911.

Kassner, M. E. and Breithaupt, R. D. (1984). The yield stress of type 21-6-9 stainless steel over a wide range of strain rate (10^{-5}–10^{4} s^{-1}) and temperature. In *Mechanical Properties at High Rates of Strain 1984*, Institute of Physics Conference Series No. 70, ed. J. Harding, pp. 47–54. Bristol: Institute of Physics.

Kim, K. S., Clifton, R. J., and Kumar, P. (1977). Combined normal-displacement and transverse-displacement interferometer with an application to Y-cut quartz. *Journal of Applied Physics*, **48**, 4132–4139.

Klepaczko, J. R. (1994). An experimental technique for shear testing at high and very high strain rates. The case of a mild steel. *International Journal of Impact Engineering*, **15**, 25–39.

Klepaczko, J. R., Lipinski, P., and Molinari, A. (1988). An analysis of the thermoplastic catastrophic shear in some metals. In *Impact Loading and Dynamical Behavior of Materials*, Vol. 2, eds. C. Y. Chiem, H. D. Kunze, and L. W. Meyer, pp. 695–704. Germany: DGM Informationsgesellschaft mbH.

Klopp, R. W., Clifton, R. J., and Shawki, T. G. (1985). Pressure-shear impact and the dynamic viscoplastic response of metals, *Mechanics of Materials*, **4**, 375–385.

Kocks, U. F., Argon, A. S., and Ashby, M. F. (1975). Thermodynamics and kinetics of slip. In *Progress in Materials Science*, Vol. 19. New York: Pergamon Press.

Kröner, E. (1968). Initial studies of a plasticity theory based upon statistical mechanics. Colloquium on Inelastic Behavior of Solids. Columbus, OH: Battelle Memorial Institute.

Kwon, Y. W. and Batra, R. C. (1988). Effect of multiple initial imperfections on the initiation and growth of adiabatic shear bands in nonpolar and dipolar materials. *International Journal of Engineering Science*, **26**, 1177–1187.

Lee, E. H. (1969). Elastic-plastic deformations at finite strain. *Journal of Applied Mechanics*, **36**, 1–6.

Lennon, A. M. and Ramesh, K. T. (1998). A technique for measuring the dynamic behavior of materials at high temperatures. *International Journal of Plasticity*, **14**, 1279–1292.

Liao, S.-C. and Duffy, J. (1998). Adiabatic shear bands in a Ti-6Al-4V titanium alloy. *Journal of the Mechanics and Physics of Solids*, **46**, 2201–2232.

Lindholm, U. S. (1964). Some experiments with the split Hopkinson pressure bar. *Journal of the Mechanics and Physics of Solids*, **12**, 317–335.

Lindholm, U. S., Nagy, A., Johnson, G. R., and Hoegfeldt, J. M. (1980). Large strain, high strain rate testing of copper. *Journal of Engineering Materials and Technology*, **102**, 376–381.

Litonski, J. (1977). Plastic flow of a tube under adiabatic torsion. *Bulletin de l'Academie Polonaise des Sciences – Serie des Sciences Techniques*, **25**, 7–14.

Lubliner, J. (1990). *Plasticity Theory*. New York: Macmillan.

Maddocks, J. H. and Malek-Madani, R. (1992). Stability of localized steady state solutions in thermoplasticity. *International Journal of Solids and Structures*, **29**, 2039–2061.

Magness, L. S. (1993). Properties and performance of KE penetrator materials. In *International Conference on Tungsten and Tungsten Alloys – 1992*, eds. A. Bose and R. J. Dowding, pp. 15–22. Princeton, NJ: Metal Powders Industries Federation.

(1994). High strain rate deformation behaviors of kinetic energy penetrator materials during ballistic impact. *Mechanics and Materials*, **17**, 147–154.

Marchand, A. and Duffy, J. (1988). An experimental study of the formation process of adiabatic shear bands in a structural steel. *Journal of the Mechanics and Physics of Solids*, **36**, 251–283.

Markus, L. and Yamabe, H. (1960). Global stability criteria for differential systems. *Osaka Mathematics Journal*, **12**, 305–317.

Mason, J. J., Rosakis, A. J., and Ravichandran, G. (1994). Full field measurements of the dynamic deformation field around a growing adiabatic shear band at the tip of a dynamically loaded crack or notch. *Journal of the Mechanics and Physics of Solids*, **42**, 1679–1697.

Merzer, A. M. (1982). Modeling of adiabatic shear band development from small imperfections. *Journal of the Mechanics and Physics of Solids*, **30**, 323–338.

Mescall, J. and Weiss, V. (eds.) (1983). Material behavior under high stress and ultrahigh loading rates. 29th Sagamore Army Materials Conference. New York: Plenum Press.

Meunier, Y., Roux, R., and Moureaud, J. (1992). Survey of adiabatic shear phenomena in armor steels with perforation. In *Shock Wave and High-Strain-Rate Phenomena in Materials*, eds. M. A. Meyers, L. E. Murr, and K. P. Staudhammer, pp. 637–644. New York: Marcel Dekker.

Meyers, M. A. (1994). *Dynamic Behavior of Materials*. New York: Wiley.

Meyers, M. A., Chen, Y.-J., Marquis, F. D. S., and Kim, D. S. (1995). High-strain, high-strain-rate behavior of tantalum. *Metallurgical and Materials Transactions*, **26A**, 2493–2501.

Meyers, M. A. and Wittman, C. L. (1990). Effect of metallurgical parameters on shear band formation in low-carbon (∼0.20 wt pct) steels. *Metallurgical Transactions*, **21A**, 3153–3164.

Miller, A. K. (ed.) (1987). *Unified Constitutive Equations for Creep and Plasticity*. London: Elsevier Applied Science.

Molinari, A. (1997). Collective behavior and spacing of adiabatic shear bands. *Journal of the Mechanics and Physics of Solids*, **45**, 1551–1575.

Molinari, A. and Clifton, R. J. (1987). Analytical characterization of shear localization in thermoviscoplastic materials. *Journal of Applied Mechanics*, **54**, 806–812.

Moss, G. L. (1981). Shear strains, strain rates and temperature changes in adiabatic shear bands. In *Shock Waves and High-Strain-Rate Phenomena in Metals*, eds. M. A. Meyers and L. E. Murr, pp. 299–312. New York: Plenum Press.

Needleman, A. and Tvergaard, V. (1995). Analysis of a brittle-ductile transition under dynamic shear loading. *International Journal of Solids and Structures*, **32**, 2571–2590.

Nemat-Nasser, S. and Isaacs, J. B. (1997). Direct measurement of isothermal flow stress of metals at elevated temperatures and high strain rates with application to Ta and Ta-W alloys. *Acta Materialia*, **45**, 907–919.

Nemat-Nasser, S., Isaacs, J. B., and Starrett, J. E. (1991). Hopkinson techniques for dynamic recovery experiments. *Proceedings of the Royal Society of London*, **A435**, 371–391.

Nemat-Nasser, S. and Li, Y. (1998). Flow stress of f.c.c. polycrystals with application to OFHC copper. *Acta Materialia*, **46**, 565–577.

Nesterenko, V. F. and Bondar, M. P. (1994). Localization of deformation in collapse of a thick-walled cylinder. *Combustion, Explosion, and Shock Waves*, **30**, 500–509.

Nesterenko, V. F., Meyers, M. A., and Wright, T. W. (1998). Self-organization in the initiation of adiabatic shear bands. *Acta Materialia*, **46**, 327–340.

Nicholas, T. (1982). Material behavior at high strain rates. Chapter 8 in *Impact Dynamics*, eds. J. A. Zukas, T. Nicholas, H. F. Swift, L. B. Greszczuk, and D. R. Curran. New York: Wiley.

Nicholas, T. and Bless, S. J. (1985). High strain rate tension testing. In *ASM Metals Handbook*, 9th ed., Vol. 8, *Mechanical Testing*, pp. 208–214. Materials Park, OH: ASM International.

National Materials Advisory Board (1980). *Materials Response to Ultra-High Loading Rates*, NMAB-356. Washington, DC: National Academy of Sciences.

Ogden, R. W. (1984). *Non-Linear Elastic Deformations*. New York: Halsted Press, Wiley.

Olson, G. B. (1997). Computational design of hierarchically structured materials. *Science*, **277**, 1237–1242.

Ortiz, M., and Popov, E. P. (1982). A statistical theory of polycrystalline plasticity. *Proceedings of the Royal Society of London*, **A379**, 439–458.

Raftenberg, M. N. (2000). A shear banding model for penetration calculations. *International Journal of Impact Engineering*, **25**, 123–146.

Rajagopol, K. R. and Srinivasa, A. R. (1998a). Mechanics of the inelastic behavior of materials – Part I, Theoretical underpinnings. *International Journal of Plasticity*, **14**, 945–967.

(1998b). Mechanics of the inelastic behavior of materials – Part II: Inelastic response. *International Journal of Plasticity*, **14**, 969–995.

Ramesh, K. T. and Narasimhan, S. (1996). Finite deformations and the dynamic measurement of radial strains in compression Kolsky bar experiments. *International Journal of Solids and Structures*, **33**, 3723–3738.

Ravi-Chandar, K. (1995). On the failure mode transitions in polycarbonate under dynamic mixed-mode loading. *International Journal of Solids and Structures*, **32**, 925–938.

Recht, R. F. (1964). Catastrophic thermoplastic shear. *Journal of Applied Mechanics, Transactions of the American Society of Mechanical Engineers*, **86**, 189–193.

Rice, J. R. (1971). Inelastic constitutive relations for solids: an internal variable theory and its application to metal plasticity. *Journal of the Mechanics and Physics of Solids*, **19**, 433–455.

Rogers, H. C. (1979). Adiabatic plastic deformation. *Annual Review of Materials Science*, **9**, 283–311.

(1983). Adiabatic shearing – general nature and material aspects. In *Material Behavior Under High Stress and Ultrahigh Loading Rates*, 29th Sagamore Conference, eds. J. Mescal and V. Weiss, pp. 101–118. New York: Plenum Press.

Rogers, H. C. and Shastry, C. V. (1981). Material factors in adiabatic shearing in steels. In *Shock Waves and High-Strain-Rate Phenomena in Metals*, eds. M. A. Meyers and L. E. Murr, pp. 285–298. New York: Plenum Press.

Rosakis, P., Rosakis, A. J., Ravichandran, G., and Hodowany, J. (2000). A thermodynamic internal variable model for the partition of plastic work into heat and stored energy in metals. *Journal of the Mechanics and Physics of Solids*, **48**, 581–607.

Sackett, S. J., Kelly, J. M., and Gillis, P. F. (1972a). A probabilistic approach to polycrystalline plasticity – Part I: Theory. *Journal of the Franklin Institute*, **304**, 33–46.

(1972b). A probabilistic approach to polycrystalline plasticity – Part II: Applications. *Journal of the Franklin Institute*, **304**, 47–63.

Schaffer, J. P., Saxena, A., Antolovich, S. D., Sanders, T. H., Jr., and Warner, S. B. (1995). *The Science and Design of Engineering Materials*. Chicago: Irwin.

Scheidler, M. J. and Wright, T. W. (2001). A continuum framework for finite viscoplasticity. *International Journal of Plasticity*, **17**, 1033–1085.

Semiatin, S. L., Lahoti, G. D., and Oh, S. I. (1983). The occurrence of shear bands in metalworking. In *Material Behavior Under High Stress and Ultrahigh Loading Rates*, 29th Sagamore Conference, eds. J. Mescal and V. Weiss, pp. 119–159. New York: Plenum Press.

Semiatin, S. L. and Jonas, J. J. (1984). *Formability and Workability of Metals: Plastic Instability and Flow Localization*. ASM Series in Metal Processing, Vol. 2, ed. H. L. Gegel. Metals Park, OH: American Society for Metals.

Senseny, P. E., Duffy, J., and Hawley, R. H. (1978). Experiments on strain rate history and temperature effects during plastic deformation of close-packed metals. *Journal of Applied Mechanics*, **45**, 60–66.

Shawki, T. G. and Clifton, R. J. (1989). Shear band formation in thermal viscoplastic materials. *Mechanics of Materials*, **8**, 13–43.

Silling, S. A. (1992). Shear band formation and self-shaping in penetrators. SAND92-2692. Albuquerque, NM: Sandia National Laboratory.

Steinberg, D. J. and Lund, C. M. (1989). A constitutive model for strain rates from 10^{-4} to 10^{6} s^{-1}. *Journal of Applied Physics*, **65**, 1528–1533.

Taylor, G. I. and Quinney, H. (1937). The latent heat remaining in a metal after cold working. *Proceedings of the Royal Society of London*, **A163**, 157–181.

Timothy, S. P. and Hutchings, I. M. (1985). The structure of adiabatic shear bands in a titanium alloy. *Acta Metallurgica*, **33**, 667–676.

Tomita, Y. (1994). Simulations of plastic instabilities in solid mechanics. *Applied Mechanics Reviews*, **47**, 171–205.

Toupin, R. A. and Rivlin, R. S. (1960). Dimensional changes in crystals caused by dislocations. *Journal of Mathematical Physics*, **1**, 8–15.

Tresca, H. (1878). On further application of flow of solids. *Proceedings of the Institute of Mechanical Engineers*, **30**, 301–345.

Truesdell, C. A. and Toupin, R. A. (1960). The classical field theories of mechanics. In *Handbuch der Physik*, Vol. III/3, ed. S. Flugge. Berlin: Springer-Verlag.

Walter, J. W. (1992). Numerical experiments on adiabatic shear band formation in one dimension. *International Journal of Plasticity*, **8**, 657–693.

Walter, J. W. and Kingman, P. W. (1998). High obliquity impact of a compact penetrator on a thin plate: penetrator splitting and adiabatic shear. ARL-TR-1584. Aberdeen Proving Ground, MD: Army Research Laboratory.

Weerasooriya, T. and Beaulieu, P. A. (1993). Effects of strain-rate on the deformation and failure behavior of 93W-5Ni-2Fe under shear loading. *Materials Science and Engineering*, **A172**, 71–78.

Wingrove, A. L. (1973). The influence of projectile geometry on adiabatic shear and target failure. *Metallurgical Transactions*, **4**, 1829–1833.

Wright, T. W. (1982). Stared energy and plastic volume change. *Mechanics of Materials*, **1**, 185–187.

(1987a). Some aspects of adiabatic shear bands. In *Metastability and Incompletely Posed Problems*, eds. S. Antman, J. L. Ericksen, D. Kinderlehrer, and I. Muller, pp. 353–372. New York: Springer-Verlag.

(1987b). Steady shearing in a viscoplastic solid. *Journal of the Mechanics and Physics of Solids*, **35**, 269–282.

(1990a). Approximate analysis for the formation of adiabatic shear bands. *Journal of the Mechanics and Physics of Solids*, **38**, 515–530.

(1990b). Adiabatic shear bands. *Applied Mechanics Reviews*, **43**, S196–S200.

(1992). Shear band susceptibility: work hardening materials. *International Journal of Plasticity*, **8**, 583–602.

(1994). Toward a defect invariant basis for susceptibility to adiabatic shear bands. *Mechanics of Materials*, **17**, 215–222.

Wright, T. W. and Batra, R. C. (1985). The initiation and growth of adiabatic shear bands. *International Journal of Plasticity*, **1**, 202–212.

Wright, T. W. and Ockendon, H. (1992). A model for fully formed shear bands. *Journal of the Mechanics and Physics of Solids*, **40**, 1217–1226.

(1996). A scaling law for the effect of inertia on the formation of adiabatic shear bands. *International Journal of Plasticity*, **12**, 927–934.

Wright, T. W. and Ravichandran, G. (1997). Canonical aspects of adiabatic shear bands. *International Journal of Plasticity*, **13**, 309–325.

Wright, T. W. and Walter, J. W. (1987). On stress collapse in adiabatic shear bands. *Journal of the Mechanics and Physics of Solids*, **35**, 701–720.

(1989). Adiabatic shear bands in one dimension. In *Mechanical Properties of Materials at High Rates of Strain 1989*, Institute of Physics Conference Series No. 102, ed. J. Harding, pp. 119–126. Bristol: Institute of Physics.

(1994). On mode III propagation of adiabatic shear bands. In *Proceedings of the Fifth International Conference On Hyperbolic Problems: Theory, Numerics, Applications*, eds. J. Glimm, M. J. Graham, J. W. Grove, and B. J. Plohr, pp. 242–248. Singapore: World Scientific.

(1996). The asymptotic structure of an adiabatic shear band in antiplane motion. *Journal of the Mechanics and Physics of Solids*, **44**, 77–97.

Wulf, G. L. (1978). The high strain rate compression of aluminium. *International Journal of Mechanical Sciences*, **20**, 609–615.

Xue, Q., Nesterenko, V. F., and Meyers, M. A. (2001). Self-organization of shear bands in stainless steels: grain size effects. In *Fundamental Issues and Applications of Shock-Wave and High-Strain-Rate Phenomena*, eds. K. P. Staudhammer, L. E. Murr, and M. A. Meyers, pp. 549–559. Amsterdam: Elsevier.

Zbib, H. M., Shawki, T. G., and Batra, R. C. (ed.) (1992). Material instabilities; selected and revised proceedings of the symposium on materials instabilities. *Applied Mechanics Reviews*, **45**(3), part 2.

Zener, C. and Hollomon, J. H. (1944). Effect of strain rate upon plastic flow of steel. *Journal of Applied Physics*, **15**, 22–32.

Zerilli, F. J. and Armstrong, R. W. (1987). Dislocation-mechanics-based constitutive relations for material dynamics calculations. *Journal of Applied Physics*, **61**, 1816–1825.

(1990). Description of tantalum deformation behavior by dislocation mechanics based constitutive relations. *Journal of Applied Physics*, **68**, 1580–1590.

Zhou, M. and Clifton, R. J. (1997). Dynamic constitutive and failure behavior of a two-phase tungsten composite. *Journal of Applied Mechanics*, **64**, 487–494.

Zhou, M., Ravichandran, G., and Rosakis, A. J. (1996). Dynamically propagating shear bands in impact-loaded prenotched plates: II – numerical simulations. *Journal of the Mechanics and Physics of Solids*, **44**, 1007–1032.

Zhou, M., Rosakis, A. J., and Ravichandran, G. (1996). Dynamically propagating shear bands in impact-loaded prenotched plates: I – experimental investigations of temperature signatures and propagation speed. *Journal of the Mechanics and Physics of Solids*, **44**, 981–1006.

Zurek, A. K. and Meyers, M. A. (1996). Microstructural aspects of dynamic failure. In *High-Pressure Shock Compression of Solids II*, eds. L. Davison, D. E. Grady, and M. Shahinpoor, pp. 25–89. New York: Springer.

Index